ROUTLEDGE LIBRARY EDITIONS:
ENVIRONMENTAL AND NATURAL
RESOURCE ECONOMICS

Volume 9

NON-RENEWABLE RESOURCES AND DISEQUILIBRIUM MACRODYNAMICS

NON-RENEWABLE RESOURCES AND DISEQUILIBRIUM MACRODYNAMICS

ROBERT MARKS

Routledge
Taylor & Francis Group
LONDON AND NEW YORK

First published in 1979 by Garland Publishing, Inc.

This edition first published in 2018
by Routledge
2 Park Square, Milton Park, Abingdon, Oxon OX14 4RN

and by Routledge
711 Third Avenue, New York, NY 10017

Routledge is an imprint of the Taylor & Francis Group, an informa business

British Library Cataloguing in Publication Data
A catalogue record for this book is available from the British Library

ISBN: 978-1-138-08283-0 (Set)
ISBN: 978-1-315-14775-8 (Set) (ebk)
ISBN: 978-1-138-09054-5 (Volume 9) (hbk)
ISBN: 978-1-138-09055-2 (Volume 9) (pbk)
ISBN: 978-1-315-10859-9 (Volume 9) (ebk)

Preface

Robert Marks's book provides a full and careful analysis of an economy in which there are three markets—for labor, for energy and for produced output —whose prices are given and, at least temporarily, fixed. In consequence, some actions are impossible, even if they are desirable and feasible in budgetary terms: at these prices, some buyers may be unable to buy as much as they would like and can afford, and some sellers may be unable to sell as much as they would like and can produce. Even so, some kind of order is possible, and the job of the economist is to describe the possibilities.

Those unsatisfied demands and frustrated supplies will no doubt put some pressure on "fixed" disequilibrium prices, and eventually they may move. They may even move in the general direction of conventional supply--equals--demand equilibrium, though we do not know that. Nevertheless, if prices adjust slowly, real economies will spend a lot of time in disequilibrium situations, with some unsatisfied buyers and sellers, and analysis like that in this book will be useful in understanding what is going on.

I think that is the case, and I thought so in the 1970s when "disequilibrium" economics of this kind captured the interest and imagination of economists, including obviously Robert Marks. I thought it was a mistake when that interest dwindled, and little or no further development occurred. Why was that? Well, prices are not fixed. Disequilibrium theory needed to be completed by a theory of slow price-- change. But that is a tall order, and even more difficult in a model world in which there are latent ("notional") demands and supplies not easily expressed. That theory has not yet appeared. In addition, the fashion in economics was swinging toward more optimistic equilibrium--based versions of macroeconomics. (Opinions differ about whether that was such a good idea.) And there may have been other reasons; lines of causation in intellectual history are not usually very clear.

In any case, here is Marks's work revived, and at a time when the energy sector of the economy carries a lot of interest. Between climate change and the need to reduce the burning of fossil fuels on one side, and the uncertain development of renewable energy sources on the other side, here at least is an economic model that aims to deal with disequilibrium in energy markets.

Professor Emeritus Robert M. Solow, Nobel Laureate,
Massachusetts Institute of Technology, Cambridge, Mass.
14 April 2017

Introduction to the 2017 re-issue

As Professor Emeritus Robert M. Solow remarks in the Preface, there are fashions in economic theory. In the 1970s and early 1980s, a number of theorists, starting with Barro and Grossman (1971), began to examine general-equilibrium models that included non- market-clearing exchange. The motivation for this was that prices do not move instantaneously from one full-employment equilibrium position to another, while trade nonetheless occurs in the meantime. As my dissertation explores, allowing economic agents to buy and sell at non-market-clearing prices (or before prices have adjusted to equilibrium, if they ever do), leads to separate regimes, characterised by whether each market is a buyers' (excess supply) or a sellers' (excess demand) market. A macro model with three markets — two inputs, labour N and resource (energy) R, and one output Y — results in eight possible regimes, as outlined in Table 3.1 in the dissertation.

An agent's behaviour in one market may be constrained by the states of the other two markets he is trading in. These spillovers mean that the comparative statics of these regimes differ, so that it is not possible for agents in a constrained market to choose their position on a choice-theoretic supply or demand function.

In a survey of New Keynesian Economics published in 1990, twelve years after this dissertation was finished, Gordon (1990) remarks that: "An interesting aspect of recent U.S. new-Keynesian research is the near-total lack of interest in the general equilibrium properties of non-market-clearing models." In the U.S. "that effort is viewed as having reached a quick dead end after the insights yielded in the pioneering work" of Barro and Grossman (1971, 1976), building on the earlier contributions of Patinkin (1965), Clower (1965), and Leijonhufvud (1968).

Gordon explains this lack of interest as the consequence of a research focus, instead, on explaining sticky wages and/or prices by combining rational expectations with maximizing behaviour at the level of the individual agent. As he puts it, "Any attempt to build a model based on irrational behaviour or sub-optimal behaviour is viewed as cheating." U.S. theorists, he says, believed that it was premature to examine the broader theoretical considerations of non-market-clearing trading before the partial equilibrium problems of sticky prices are solved. Another fashion?

Forty years later, the profession understands, from behavioural economics, that irrational expectations and non-optimal behaviour are widespread, and partial equilibrium models incorporating these are emerging. But the results from the work on non-market- clearing exchange from forty years ago has not been revisited and insights from this work have been lost; no general-equilibrium models, such as the model presented in this work, have been developed recently.

Following Barro and Grossman's work, the line of research evolved in the hands of Malinvaud (1977), Mueller and Portes (1978), Benassy (1975), Grandmont (1982) and Marks (1979, 1983). Almost all of these researchers are Europeans, even if they studied at U.S. universities. But in treating this line of research with disdain (in Gordon's words), and instead focussing on the "micro foundations models as the prerequisite for macro discourse," U.S. theoreticians have, argues Gordon, overlooked the central message of the non-market-clearing trade models, which is that the failure of one market to clear imposes spillover constraints on agents in other markets.

For example, when firms in a recession experience a fall in sales at the going price, this excess supply of output spills over into a fall in labour demanded at the going real wage and a fall in resource (energy) demanded at the going real price of resource (energy). (Assuming zero short-run elasticity of substitution of resource for labour in production.)

In such a model, agents are not in a position to choose the amount they work or produce as output varies over the business cycle, and so the constrained amount that they do work or produce cannot be interpreted as tracing movements along a choice-theoretic labour supply curve or production function. This also holds for the suppliers of resource in our model with three markets.

Traditional theory holds that prices adjust quickly to excess supplies or demands, resulting in the rapid disappearance of any disequilibrium. But Leijonhufvud [1968] and Malinvaud [1977] questioned the adequacy of this theory in describing the short-run behaviour of modern market economies. The work below is my contribution to studies on the consequences of relaxing the assumption of rapid price adjustment.

The model includes three markets (for output, labour, and resource flow), with the assumption that quantity adjustment in each market in response to unbalanced supply and demand is much more rapid than price adjustment: in his survey of temporary general equilibrium theory, Grandmont (1982) characterises this kind of model as an example of "temporary equilibrium with quantity rationing," since adjustments take place in every period at least partially by quantity rationing. (Solow and Stiglitz [1968] describe a model in which quantity and price adjustments occur at comparable speeds.) In Chapter 3, we do not consider price adjustment, but treat prices as given: the speed of adjustment of prices in response to excess demand or supply can be thought of as being imperceptible in the period under analysis. (The analysis resembles that of the "fix-price" method of Hicks' [1965].)

The purpose of this model was to develop a "quasi-equilibrium" where real prices were constant, while nominal prices changed, in order to model a market for non- renewable (exhaustible) energy — such as oil. The Hotelling criterion (Hotelling 1931) was another fashion in economic theory, overtaken perhaps by concern about the finite nature of the natural environment to absorb the by-products of the combustion of fossil fuels for energy.

Clower [1965] and Barro and Grossman [1971, 1976] built models which relax the assumption of market-clearing exchange, that the amount supplied or demanded ex ante by each economic agent at the going price in each market equals ex post the actual amount traded. Exchange can occur at "false," or non-market-clearing prices. This relaxation means, first, that quantities traded cannot be determined simply by reference to market-clearing conditions (rather, the actual trading process must be examined), and, second, that agents will in general be constrained in any market by conditions they experience in other markets: their demand (and supply) functions will no longer be unconstrained, notional schedules, but will be constrained, effective schedules (Clower [1965]), and quantities will be rationed.

There is no reason to expect that the effective schedules of any agent constrained in different markets will be mutually consistent: in an economy with rationing, ex ante supplies and demands are tentative, and it is no longer optimal for the agent to determine all his schedules at a stroke. Following Benassy [1975], we let the effective demand (supply) schedule of an agent in a market be the demand (supply) he will choose by maximizing his expected utility or profit subject to his budget constraint and to the quantity constraints he perceives in the other markets: he does not take into account any constraints he might experience in the market considered.

There is thus a coordination problem: in aggregating individual schedules, we need to build a model in which there is consistency among individual actions. Malinvaud [1977] argues that there are three general properties necessary for the existence of quasi- equilibrium, in which for the given real prices quantities have no further tendency to move. First, trades balance: for each good the sum of purchases equals the sum of sales. Second, there is no involuntary exchange: no agent is forced to buy more than he demands or to sell more than he is willing to supply. Given the second property, an agent will be in one of four mutually exclusive states in a market: he will be a constrained (unconstrained) buyer if his demand exceeds (equals) his purchases; he will be a constrained (unconstrained) seller if his supply exceeds (equals) his sales. Third, there cannot exist both a constrained buyer and a constrained seller in the same market, for, were this the case, each would be able to make an advantageous trade. That is, there is one and only one market for each commodity, and all agents have free access to this market.

Given these three properties, the target amount traded in any market will be determined by the "short" side of the market (that is, it will equal the lesser of the amounts supplied and demanded), and agents on the "long" side of the market will be constrained in their transactions, implying some means of rationing. The market for any commodity is then in one of three states: it can be balanced (with clearing and no rationing), or a sellers' market (with constrained buyers), or a buyers' market (with constrained sellers). We assume that the pattern of rationing does not affect the aggregate levels of the effective demands and supplies in the economy. (With this assumption and those of fixed supply of labour and of resource flow, we sidestep the conclusions of Hildenbrand and Hildenbrand [1978] that there is no sound foundation for the non- market-clearing comparative statics propositions derived by Malinvaud [1977].)

We assume that there is no inventory accumulation. (Blinder [1981] and Green and Laffont [1981] discussed the implications of this for non-market-clearing analysis.) Further, we assume that costs of quantity adjustment are zero, which excludes the possibility of levels of output or inputs independent of prices or sales: firms set output to be equal to sales at all times and minimize the costs of the input factors given this level of output.

There are different responses in the level of employment across the regimes. From Table 3.5 we see that a rise in the real resource (energy) price will tend to decrease employment in the regime SC (Malinvaud's "classical unemployment"), but will tend to increase employment in the regime DC (Malinvaud's "Keynesian unemployment") (at least for Cobb-Douglas technology); it will not affect employment in any other regime. (See Table 3.1 for the regime definitions.) (Malinvaud [1977] claims that this distinction was responsible for much confusion in the policy debates of the 'thirties.) In an extension of Chapter 3, Marks (1983, Table 3) shows that a fall in resource (energy) supply will tend to reduce employment in regime RC, to increase it in regime DRC, while not affecting it in other regimes; and a fall in autonomous demand for output will tend to reduce employment in regimes DC and DRC, but will not affect it in other regimes.

In Chapter 4, the dissertation does allow nominal prices to respond to unbalanced supply and demand in a closed economy, by extending the model to include Walrasian price adjustment using two possible formulations; Solow [1980] does this for an economy with completely elastic resource supply. In Chapter 5, we explore expectations of prices, the supply of resources (energy), and the Hotelling principle.

In a paper examining the implications of different assumptions concerning the relative speeds of price and quantity adjustment in the output and labour markets, Corden [1978] attempts to allocate "responsibility" for unemployment—whether the government or households (through the autonomous demand for output), or "big business" (through the price of output), or trade unions (through the wage). In an analogous manner we could ascribe unemployment in, say, the SC regime of classical unemployment to the cost of input factors: if either the real wage or the real resource price fell, output and employment would increase; a fall of the real wage in regimes DC (of

Keynesian unemployment) and RC would likewise increase employment. But it is difficult in our model, with two variable input factors, to ascribe "responsibility" for unemployment to any single group. Rather, the regime in which the economy finds itself is a function of the supplies and real prices of resource and labour, the exogenous demand for output, and the degree of leakage of aggregate demand.

Acknowledgments

I should like to acknowledge my gratitude to Bob Solow for kindly writing a preface for this republication, and to thank Geoff Harcourt, Chris Adam, Peter Saunders, Joe Stiglitz, Duncan Foley, the late George Danzig, and my supervisor Jim Sweeney (who was instrumental in the 1979 publication of this dissertation by Garland Publishing, New York).

Robert E. Marks
2 May 2017

References

Barro, R. J., and H. I. Grossman [1971], A general disequilibrium model of income and employment, *American Economic Review*, 61: 82–93.

Barro, R. J., and H. I. Grossman [1976], *Money, employment, and inflation,* New York: Cambridge University Press.

Benassy, J.-P. [1975], Neo-Keynesian disequilibrium theory in a monetary economy, Review of Economic Studies, 42: 503–523.

Blinder, A. S. [1981], Inventories and the structure of macro models, *American Economic Review*, 71: 11–16.

Clower, R. W. [1965], The Keynesian counterrevolution, in F. Hahn and F. Brechling (eds), *The theory of interest rates*, London: Macmillan. Reprinted in R. W. Clower, ed., *Monetary Theory: Selected Readings,* Harmondsworth: Penguin, 1969.

Corden, W. M. [1978], Keynes and the others: wage and price rigidities in macroeconomic models, *Oxford Economic Papers*, 30: 159–180.

Gordon, R. J. [1990], What Is New-Keynesian Economics? *Journal of Economic Literature*, 28: 1115–1171.

Grandmont, J.-M. [1982], Temporary general equilibrium theory, *Handbook of Mathematical Economics*, vol. II, edited by K. J. Arrow and M. D. Intriligator, North-Holland Publishing, Chapter 19, pp. 879–922.

Grandmont, J.-M., and G. Laroque, [1977], On temporary Keynesian equilibrium, in The Microeconomic Foundations of Macroeconomics: Proceedings of a Conference held by the International Economic Association at S'Agaro, Spain, pp. 41–61, ed. by G.

C. Harcourt, London: Palgrave Macmillan.

Green, J., and J.-J. Laffont [1981], Disequilibrium dynamics with inventories and anticipatory price-setting, *European Economic Review*, 16: 199–221.

Hicks, J. [1965], *Capital and growth.* Oxford: Clarendon Press.

Hildenbrand, K., and W. Hildenbrand [1978], On Keynesian equilibrium with unemployment and quantity rationing, *Journal of Economic Theory*, 18: 255–277.

Hotelling, H. [1931], The economics of exhaustible resources, *Journal of Political Economy*, 39: 137–175.

Leijonhufvud, A. [1968], *On Keynesian economics and the economics of Keynes,* New York: Oxford University Press.

Malinvaud, E. [1977], *The theory of unemployment reconsidered*, Oxford: Basil Blackwell.

Marks, R. E. [1979], *Disequilibrium macrodynamics and non-renewable resources,* New York: Garland Publishing. [This dissertation is here reprinted, with prefaces.]

Marks, R. E. [1983], Energy, output, and employment in the short run, (presented at the 1979 Econometric Society Meetings, Atlanta), Australian Graduate School of Management Working Paper 83-030, HYPERLINK "http://www.agsm.edu.au/bobm/papers/agsm-wp-83-030.pdf" \h http://www.agsm.edu.au/bobm/papers/agsm-wp-83-030.pdf

Muellbauer, J. and R. Portes [1978], Macroeconomic models with quantity rationing, Economic Journal, 88: 788–821.

Patinkin, D. [1965], *Money, interest and prices*, 2nd edn. New York: Harper and Row.

Solow, R. M. [1980], What to do (macroeconomically) when OPEC comes, in S. Fisher (ed.), *Rational expectations and economic policy*, Chicago: University of Chicago Press for the National Bureau of Economic Research.

Solow, R. M., and J. E. Stiglitz [1968], Output, employment, and wages in the short run,Quarterly Journal of Economics, 82: 537–560.

Non-Renewable Resources and Disequilibrium Macrodynamics

Robert Marks

Garland Publishing, Inc.
New York & London, 1979

Library of Congress Cataloging in Publication Data

Marks, Robert, 1946–
Non-renewable resources and disequilibrium macrodynamics.

(Outstanding dissertations on energy) (Outstanding dissertations in economics)
Originally presented as the author's thesis, Stanford, 1978.
Bibliography: p.
Includes index.
1. Power resources—Mathematical models.
2. Natural resources—Mathematical models.
3. Equilibrium (Economics) I. Title. II. Series.
III. Series: Outstanding dissertations in economics.
HD9502.A2M35 1979 333.7 78-75018
ISBN 0-8240-4053-8

All volumes in this series are printed on acid-free, 250-year-life paper.

Printed in the United States of America.

To my parents,
Joan and Leslie.

ABSTRACT

Recent research on temporary equilibria with quantity rationing has relaxed the assumption of adjustments occurring only through price movements and has examined the implications of "sticky" prices and adjustments occurring through quantity rationing for the explanation of such real-world phenomena as involuntary unemployment. This study extends the earlier work by examining a simple fix-price model with three goods: output, labour, and another factor input, which will be thought of as a flow of raw materials. This is shown to lead to the possibility of eight regions of partial market clearing in the real price plane. The appropriate fiscal policy to alleviate unemployment or to stimulate production will vary depending on the relevant region.

The study continues by examining the question of whether a competitive economy can efficiency allocate a stock of non-renewable natural resources through time. Long-run analyses of competitive economies with such resources have concluded that, without perfect foresight or a complete set of futures markets extending infinitely far into the future, there is no economic mechanism to guarantee that the initial price is set so that the economy converges to the socially desirable path of balanced growth.

But in order to reach this conclusion, the authors of the long-run analyses have made very strong assumptions, in particular that markets clear instantaneously and that arbitrage between resource and asset markets occurs so rapidly that the rate of return on holding stocks of resource equals the return on holding other assets. This study relaxes both assumptions. It examines the existence, uniqueness, and stability of the short-run equilibrium of a simple, four-market, competitive economy, including stocks of non-renewable natural resources.

Analysis of the interdependent short-run adjustments of the markets as prices react to perceived excess demands cannot be made if market clearing is already assumed. Moreover, the behaviour of the participants in the resource stock/asset market might lead to equality of rates of return, but the question cannot be answered by assumption. A further relaxation in this study will be to examine modes of expectation formation: the only way a stock of non-renewable natural resource can produce a current return for its owner is by appreciating in value; hence the anticipated price plays an important role in the behaviour of participants in the resource stock market.

As well as the resource stock/asset market, the study includes a flow market for resources, a labour market, and a market for output, a system of markets linked by the profit-maximizing representative firm, which buys labour services and flow of resource as factor inputs in the short-run, and sells its produced output to the households, which as well as earning the wage bill are assumed to receive the net profits from industry and the return to owners of resource supply on the resource market.

Our analysis indicates that equilibrium of the three flow markets cannot exist with equilibrium of the stock/asset market for resources. Stable equilibrium can occur only with constant real resource flow price. Thus the analysis indicates that the long-run growth paths, even if eventually stable, are not supported by the micro-behaviour of the system in the short-run. This implies that it is not sufficient for efficient allocation of non-renewable natural resources that participants have perfect foresight or that a complete set of futures markets exist, since, although any long-run growth path would then be stable in the long-run sense, it would remain unstable in the short-run.

ACKNOWLEDGEMENTS

I am deeply indebted to Professor James Sweeney for his guidance during the course of the dissertation's writing. His detailed criticisms and suggestions, and many spirited discussions with him, have led to many insights and have greatly improved the explication of the study. Professor Karen Johnson has been most generous with her time in reading the manuscript and discussing areas in which the study could be improved. I wish to thank Professor Edison Tsé for his valuable assistance and comments.

I wish to thank, too, Professor George Danzig, who chaired my orals committee, and Professor Willis Harman, who represented the Engineering-Economic Systems department at the oral examination, and who reminded me of the economist's moral responsibilities.

This field of study was first suggested to me by Professor Joseph Stiglitz, and I thank him for his encouragement in the early stages. I also wish to thank Professors Duncan Foley, David Starrett, Pentti Kouri, Lawrence Lau, and Bert Hickman for their helpful comments and suggestions.

I wish to thank Claire Gilchrist for her fine and tireless typing, and Professor Sherman Maisel and the National Bureau of Economic Research-West for letting me use their facilities. I wish to thank Professor Philip Brown and the Australian Graduate School of Management for special leave to complete the dissertation.

This research was partially supported by a Commonwealth Scientific and Industrial Research Organisation of Australia Scholarship, a Resources for the Future Fellowship, and the U.S. Energy Research and Development Administration Grant ERDA 1488.

TABLE OF CONTENTS

LIST OF TABLES

LIST OF FIGURES

List of Figures, continued

CHAPTER I: INTRODUCTION

1.1. The economics of non-renewable natural resources.

In the past few years there has been an increasing awareness of the finite "limits" of the "Spaceship Earth." In economics this has shown itself in a renewed interest in the role of natural resources in the economic process, and the effects of the institutions and level of activity of the economy in the conservation and use of natural resources. This study is concerned with both of these aspects of the economics of non-renewable natural resources. In particular, does the existence of non-renewable natural resources in the economy have any short-run effects on the interactions of the economy--does their existence affect short-run stabilisation policies, and, if so, how? When we consider the short-run micro-foundations of economic behaviour is there any reason to believe that the competitive economy will exploit the resources at the socially desirable rate? We hope to show that the role of natural resources in the economy does lead to alterations in the conventional, Keynesian, policy prescriptions for unemployment. We hope to show that, in the short run, a necessary (but not sufficient) condition for the efficient intertemporal allocation of non-renewable natural resource, the Hotelling principle, will not be satisfied, and that the economy will tend towards a state which cannot be intertemporally optimal.

Almost all of the theoretical economic analyses of resource economics have been micro-economic in nature: in the past four years there have been many efforts to extend and apply the principles of micro-economic theory to the discovery, extraction, and utilization of natural resources. In particular there have been several studies on optimal growth in an economy including exhaustible natural resources. In such a study, the long-run stability has been examined by Stiglitz (1974b), who considers an economy moving along an equilibrium path along which expectations about future prices are realized and along which markets clear at every moment. He shows that without a complete set of futures markets extending infinitely into the future there is no economic mechanism to guarantee that the initial price will be set so that the economy converges to balanced growth: if the initial price of natural resource is set too low the resource stock will be used up in a finite time; if the initial price is set too high there is always a finite amount of resource stock remaining--an inefficient situation with over-saving of resource. Stiglitz also shows that, unlike the heterogeneous capital growth model, the natural resource model exhibits long-run instability with even slow rates of adaptive expectation formation of the rate of change of resource price.

But this analysis is made with the assumption that the economy is in equilibrium with markets clearing at every "moment," where each moment is vanishingly short so that a continuous analysis can be made. In particular, the explicit assumption is made that the asset market (including stocks of non-renewable natural resource) is in equilibrium so that the return on holding stocks of natural resources (the proportional change of resource price, net of extraction costs) is equal to the rate of return on other assets (the interest rate). This, briefly stated, is the Hotelling principle. If all markets clear, what leads to the change of resource price? Traditionally we have considered prices to change in response to an imbalance of supply and demand. We shall look more closely at this question in Chapter V.

Short-run stability is concerned with perturbations around the long-run path: if the economy suffers a shock exogenous to the model (such as a change in price expectation formation, or a change in household saving behaviour, or a new discovery of resource, or a change in government policy, or a change in production technology), will equilibrium (in the sense of markets clearing with expectations realized) be regained, and if so will the previous equilibrium path be regained? To answer these questions the short-run equilibrium must be analyzed for existence, uniqueness, and stability.

Stiglitz (1974b) suggests that it is possible in the short run that disequilibrium in the natural resource market (stemming from exogenous disturbances of the supply of resource, for instance) may be translated into disequilibria in other markets (output, asset, and labour). The consequent adjustments in these markets would affect the level of employment and the wage rate, the supply of output and the price level, and the rate of investment and the interest rate.

The linkage between the natural resource market and the asset market could be the cause of a knife-edge instability. With perfectly competitive markets we should expect the price of the flow of resources to equal the price of the stock of resources. The knife-edge occurs when people's expectations of the rate change of the resource price change: if the expected proportional rate of change of resource price exceeds the interest rate, and if this situation is expected to continue, then traders in the asset market will attempt to increase their stocks of resources. This increase in the demand for stocks will lead to a greater increase in the resource price than would have occurred before, and since traders' expectations are fulfilled they will continue to view resource stocks as a better investment than other assets. At the same time there may be an effect downwards on the interest rate: the increase in resource price will lead to a substitution away from resource flow as a factor input, towards men and machines, which will lead to a change in the utilization of machines and hence to a change in the marginal and average products of capital, in turn leading to a change in the interest rate. If the final effect on the interest rate is downwards then there is even stronger stimulus to hold stocks of resources, and the disequilibrium is exacerbated--hence the knife-edge.

This problem is similar to other speculative "bubbles," and is similar too to the Hahn problem which occurs with heterogeneous capital goods: if the price of one capital good were initially set "too high," the price of the asset would have to rise faster, for market-clearing of the particular capital good, than it would if the price were lower, to offset the lower value of the rentals per dollar invested. Thus in the next period the price would be even further "out of line." One motivation for the following work has been to attempt to build an analytical model to study this behaviour. Insight into possible linkages in the economy between the supply schedule of the natural resource and the adjustments of the markets for output and labour will lead to more effective policy prescriptions; and the work will more fully integrate natural resources into short-run macroeconomic theory.

Since the analysis is concerned with disequilibrium interactions among such macroeconomic variables as output, employment, and inflation, the final model will include four markets: a market for output, a market for labour services, a market for the resource, and a market for assets. Corresponding to these four markets there will be four prices: a money price of output, a money wage, a money price of resource, and an interest rate.

1.2. Disequilibrium adjustment with non-market-clearing trading.

In order to examine the interaction of the markets during the disequilibrium adjustment, the assumption of no non-market-clearing trading will be relaxed. This traditional assumption has been explained by two equivalent descriptions of price determination: either that prices adjust to excess demands instantaneously, or that prices are determined by a Walrasian tatonnement process (or Edgeworthian recontracting process) in which no production or exchange occur until the equilibrium price vector is reached. The alternative assumption will be made that prices do not adjust instantaneously, but that production and exchange can occur at "false" (that is, non-market-clearing) prices.

As Hicks (1946) argues, trading at "false" prices leads to income effects, which will only be negligible if all traders' expenditures on a good are only a small part of their total incomes, with the market ending up very close to the equilibrium price. But in an aggregate macro model these expenditures will not be small in relation to each economic actor's total income, and the income effects from trading at "false" prices will affect the final market price, and will have to be included in the model.

Realization of this leads to Clower's (1965) "dual decision hypothesis" formulation of "income-constrained" processes, in which income as well as prices is an argument in each trader's demand (supply) function. Clower speaks of "notional" demand and supply schedules with no quantity constraints (no non-market-clearing trading), and "effective" schedules which include the quantity constraints which result from non-market-clearing trading. (The Keynesian aggregate demand function is an example of an effective demand schedule: income, as well as prices, is an argument.) Leijonhufvud (1968) stresses that income-constrained processes result whenever the speed of price adjustment is less than infinite. Thus analysis of disequilibrium adjustment requires both price and quantity adjustments.

Four markets lead to four prices and four quantities. A general disequilibrium analysis would involve explicit adjustments in these eight variables. In fact Solow and Stiglitz (1968) describe a disequilibrium macro model with explicit adjustments for employment, the money wage, and the money price of output. They define short-run equilibrium as occurring when the level of employment and the real wage rate are constant, a definition which includes not only the conventional equilibrium with all variables constant, but also Hansen's (1951) "quasi-equilibrium" in which real prices are constant but money prices are changing at equal proportional rates, with (positive or negative) excess demand resulting from lack of market clearing. But the technical problems of eight separate adjustment processes are great, and it is possible to reduce the number of independent adjustments without destroying the essential disequilibrium nature of the model.

The complexity of continuous adjustment models is a great incentive to adopt a discrete treatment of time into "periods," which requires a qualitative ranking of the adjustment speeds of the variables in the system. It is then convenient to treat variables adjusting relatively slowly as data, and to treat variables adjusting relatively rapidly as having worked out their effects. Thus the explicit adjustment processes are reduced to those of interest only. The model to be developed below makes the assumption that quantities adjust infinitely faster than prices, which Leijonhufvud (1968) asserts is the "revolutionary" element in Keynes' General Theory. With this assumption, in an ultra-short-run period or "momentary situation," all prices are given, and on the basis of these prices all plans are formulated and quantities adjust on each market according to derived rules. If at the given prices there are non-zero excess demands, then price changes will be generated at the transition from one momentary situation to the next, in which the new set of prices will lead to formulation of a new set of plans and a new set of quantities. Following Korliras (1973) and Benassy (1973), in the "short-run" period of explicit analysis price changes are accompanied by instantaneous adjustments in quantities and plans which can thus be derived from current prices.

The earliest disequilibrium models were developed by Patinkin and Clower. Patinkin (1965) analyses the demand for labour in the situation where there is an excess supply of output: demand for labour is reduced and the possibility of involuntary unemployment is introduced, associated with excess (effective) labour supply. Clower (1965) analyses the other side of the coin: the demand for output and the demand for money balances subject to an employment constraint to derive effective demand functions of the same form as the usual Keynesian consumption and saving functions. These two complementary models have been brought together by Barro and Grossman (1971) and (1976).

At a higher level of abstraction, Benassy (1973) has formalized disequilibrium economics to the level of the Arrow-Debreu model of equilibrium economics. He makes the "Keynesian" assumption of instantaneous quantity adjustment and relatively slow price adjustment, and defines a "K-equilibrium" for a set of prices as a situation in which quantities have no tendency to move, more precisely a set of self-reproducing effective demands, which generate "perceived" constraints which in turn will generate the original set of effective demands. A simplified Keynesian model is built to examine the stagflation, deflation, and inflation. Price movements in response to effective excess demands follow with an examination of long-run dynamics and the Phillips curve. Benassy extends the analysis with a chapter on monopolistic competition and finally an economy with an uncertain future and where money links successive equilibria only as a store of value.

1.3. The basic assumptions.

In order to analyse the short-run disequilibrium adjustments of economy including non-renewable natural resource, we have formulated a basic model. The model includes three goods: labour services, a homogeneous output, and resource. There are three types of economic actor: firms, households, and resource suppliers. The labour services and resources are the two variable inputs in the production process: other inputs are fixed in the short-run, and have no alternative use and zero user cost. All current output is produced by the same technology, and can be considered to assume its specific identity according to the buyer: firms buy investment goods and households buy consumables. Monetary factors are ignored: it is assumed that the monetary authorities manage to keep the real rate of interest constant, and that the monetary variables do not influence aggregate demand. Firms demand labour and resource and supply homogeneous output. Firms are assumed to maximize profits, which can be considered as a return to non-variable inputs, of which each firm possesses a predetermined and fixed amount. Households supply labour. They demand consumables from firms and savings balances. They receive income from the sale of labour services, from the sale of resources, and from profits, all of which accrue only to households. It is assumed that the household decision to save can be characterized by the standard assumption of constant and equal marginal and average propensities to save out of income. The government taxes households' gross income and buys homogeneous output. In the basic model, resource flow and labour services are treated symmetrically.

The basic model is first analysed for the case in which no non-market-clearing trading occurs, that is, for the recontracting or tatonnement case. The analysis, following that of Barro and Grossman (1976) for a single variable factor input, considers comparative statics and dynamic stability of the market-clearing equilibrium that occurs with notional supply and demand schedules. In contrast the model is then analysed for the case of non-market-clearing trading, in which trading occurs during the adjustment process and effective supply and demand schedules result.

Dropping the assumption of trading only at market clearing has two essential implications for the determination of the quantities supplied and demanded. First, the quantities traded cannot be determined simply with reference to market-clearing conditions. There is no equivalence between actual transactions and the quantities supplied and demanded. In order to analyse quantity determination under non-market-clearing conditions, the actual trading process must be examined. The model includes the assumption of "voluntary exchange": no economic actor can be forced to buy more than he demands or sell more than he supplies. Consequently, the actual level of total transactions will be determined by the "short" side of the market (that is, by suppliers if there is excess demand, by demanders if excess supply), and economic actors on the "long" side will be constrained in their transactions.

These constraints lead to the second implication: not every economic actor will generally act as if he can buy or sell any amount which he demands or supplies at the existing price vector. In particular, economic actors on the "long" side of the market (that is, suppliers if excess supply, demanders if excess demand) will face quantity constraints on their transactions, to be taken into account when formulating their supplies and demands on other markets, leading to effective schedules. The notional schedules derived in a market-clearing model do not in general describe the economic behaviour of firms and households in a disequilibrium model (unless the economic actor finds himself on the "short" side of every market). The economic actor must derive his demand (supply) functions taking into account fully the information about the other markets.

In this model with three markets and instantaneous quantity adjustments, two state variables can be derived: the real wage (w) and the real resource price (v). Analysis of the model at all points in the positive quadrant of the (w, v) plane leads to eight regions in which various combinations of markets clear. These eight regions can be shown to correspond to four categories of Keynesian short-run equilibria. Thus macro behaviour has been derived from a model with profit-maximizing representative firms. The next step is to allow the prices in the three markets to adjust, using some function of the excess demands on the markets to determine the possibility of equilibria or quasi-equilibria. These can then be analyzed for stability. The models can be analysed for the comparative statics of the equilibria as the system is shocked by the changes in exogenous parameters. The model treats resource flow and labour services symmetrically.

An extension of the basic model allows the non-renewable nature of the natural resource to be included. A fourth market, that for assets, is introduced, on which stocks of non-renewable resource can be traded. The Hotelling principle, with the proportional rate of change of resource price rising at the rate of return of other assets, the interest rate, is a description of equilibrium on the asset market. We do not analyse the asset market explicitly, but assume that resource owners attempt to maximize the present value of a portfolio which includes stocks of non-renewable natural resource and other assets. Their willingness to sell these stocks as a flow of resource to the firms is assumed to be a non-increasing function of their expectation of the proportional rate of change of the resource price. Various possible formulations of this price change expectation are considered in analysis of the micro-foundations for the Hotelling principle.

1.4. Outline of the study.

In Chapter II we state the linkages of the basic model, especially including the representative behaviour of the firms, households, and resource suppliers. The market-clearing conditions are determined, and the comparative statics and dynamic stability of the basic model analysed.

In Chapter III we relax the assumption of no non-market-clearing trading, and, with prices fixed, derive the eight possible regions which are a consequence of the effective schedules. These regions are separated in the (w, v)-plane by the loci of effective market clearing, which are plotted. The comparative statics of the loci are analysed.

In Chapter IV we present three price adjustment formulations, and characterize the quasi-equilibria which result. The dynamic stability and comparative statics of the quasi-equilibria are determined.

In Chapter V we introduce the non-renewable nature of the resource to the model. The derivation of the Hotelling principle is discussed. Two resource flow supply functions and four modes of expectation formation lead to eight possible combinations. For each of these the local dynamic stability of the three possible quasi-equilibria is determined. A final stability analysis includes formulation of price adjustment which includes a direct expectational effect.

In Chapter VI we review critically the important characteristics of the models studied and the conclusions derived in the course of the study. We mention how some of our assumptions could be relaxed to strengthen the models and the possible effects of doing so. An attempt is made to consider the behaviour of the model in the longer run. Finally, we discuss some implications for future economic policy decisions from our conclusions.

CHAPTER II: THE BASIC MODEL

2.1. Essentials of the basic model

The model includes three goods: labour services, a homogeneous output of production, and a non-renewable natural resource. There are three types of decision-makers: firms, households, and resource suppliers. The labour services and flow of resource are the two variable inputs in the production process which produces the homogeneous output. Other inputs are fixed in the short-run period under consideration, and have no alternative use and zero user cost.

The representative firm is assumed to maximize profits, which can be considered as the return to the non-variable inputs. The firm demands labour services and resource flow and supplies homogeneous output. All current output is produced by the same technology, and can be considered to assume its specific identity according to the buyer: firms buy investment goods, households buy consumables, and the government buys government-demanded goods and services.

Households supply labour services. They demand consumables from firms, and savings balances. They receive income from the sale of labour services, and are assumed to receive the total amounts of profits and receipts of sales of resource flow. Monetary factors are ignored: the household demand for consumables is assumed not to depend on the rate of interest or on cash balances held by the household. It is assumed that the household decision to save can be characterized by the standard assumption of constant and equal marginal and average propensities to save out of income.

Resource suppliers are assumed to attempt to maximize the returns to their portfolios of assets, which include stocks of non-renewable natural resource. The only way such stocks can produce current returns for their owners is by appreciating in value. Hence resource suppliers compare the anticipated capital gain in value of stocks of resource held with the return from holding other assets, which throughout is assumed constant and equal to the interest rate: they are assumed to be responsive to the anticipated proportional rate of change of resource price: they will offer more (less) for sale the lower (higher) the anticipated proportional rate of price change, which is for the moment assumed to be held uniformly. Thus the net supply of resource flow will vary with expected proportional rate of price change. It is this net supply which partly or wholly satisfies the demand of the firms for resource flow as a factor input in production. For the moment resource suppliers are assumed to have identical, static expectations of the proportional rate of resource price change, and in consequence there is a positive, inelastic flow of resource for sale.

The analysis below will be concerned, at a general level, with the determination of the quantities of the goods traded and their relative prices. The distinction will be made of quantities traded, quantities demanded, and quantities supplied. The superscripts D and S will denote quantity demanded and quantity supplied, respectively, and absence of a superscript will imply the quantity actually traded. For the most part quantities will be flow variables with dimensions of stock per unit time.

Thus in the basic model the variables are:

N: flow of labour services

R: flow of resource

C: flow of consumables

T: flow of tax revenues (net of transfers)

G: flow of government purchase of goods and service

S: flow of savings

Y: flow of current output

I: flow of investment goods

In general, the number of independent variables is one less than the number of economic goods. Because consumables and investment goods are produced by the same technology, they are perfectly substitutable on the supply side, which effectively fixes the relative prices among consumables and investment goods. Thus the basic model involves three independent exchange ratios:

P: the money price of output

W: the money wage rate

V: the money price of resource flow.

The money prices have the dimensions of dollars per unit, and are also spoken of as nominal prices. Real prices are expressed in terms of output, not money, and the basic model includes:

w: the real wage rate

v: the real resource flow price

where

$$w \equiv W/P, \tag{2.1}$$

$$v \equiv V/P.$$

In analyzing the behaviour of firms, it is convenient to consider the "representative" firm, a unit whose behaviour, except for its scale, is identical with the behaviour of the aggregate of such units. The representative unit can be thought of as an average unit. It is thus possible to consider the individual and the aggregate, using the same notation to represent both. The tradeoff of this approach is that distributional effects are abstracted from. The validity of the analysis thus depends on the negligible size of any changes of distribution, or on the negligible impact of distribution variables on variables of interest in the model.

It is assumed in the basic model that firms, households, and resource suppliers are atomistic units, that is, that the marginal effect of a single unit of each in supply or demand for each good is negligible. Hence each unit neglects the impact of its own supply or demand on market price and quantity. Thus firms, households, and resource suppliers are price-takers with respect to the three nominal prices P, W, and V.

In the basic model it is assumed that each unit behaves as though it can buy or sell any amount which it demands or supplies of each good at the going market price or wage. In particular, firms believe that they can buy the quantities of labour services and natural resources which they demand and that they can sell the quantities of output which they supply; households believe that they can sell the quantity of labour services which they supply and buy the quantities of consumeables which they demand; and resource owners believe that they can sell the quantity of natural resources which they supply.

This assumption is justified *ex post* only when aggregate supply and demand balance in each market, that is, when

$$N^D = N^S = N$$

and

$$R^D = R^S = R$$

and

$$C^D = C^S = C = Y - (I + G)$$

since

$$Y^D \equiv C^D + I + G = Y^S \equiv C^S + I + G = Y.$$

In traditional models exchanges are assumed to take place only under these market-clearing conditions, under which each atomistic unit's *ex ante* assumption of the ability to buy or sell any amount desired at the existing P, W, and V is always borne out *ex post* by the actual transactions which occur in the markets. Following Clower (1965) the demand and supply functions derived for the basic model, in which market-clearing is assumed to be able to occur at the going P, W, and V with no constraints on purchases or sales, are called "notional" functions.

If non-market-clearing exchange is considered, then the possibility is introduced that actual transactions and quantities supplied or demanded might be unequal for some economic units. In this case, economic decision-makers in formulating supply and demand schedules might consider possible constraints on purchases or sales. These general demand and supply functions are called "effective" functions (Clower (1965)).

2.2. Representative behaviour in the basic model.

2.2.1. The firms.

It is assumed that the object of firms is to maximize profits. Since households are assumed to receive profits, this objective can be thought of as being imposed on the firms by the households. Given the profit-maximizing objective, firms can be viewed as decision-making units distinct from households.

Profits are the difference between sales revenues and payments to factor inputs. The basic model assumes that the only inputs which must be bought are labour services and resource flow. Hence, in real terms, profit (π) is given by

$$\pi = Y - wN - vR. \tag{2.2}$$

With the assumption that firms can buy the quantities of labour services and natural resources they demand, and can sell the quantity of output they supply, profit as viewed by the representative firm may also be expressed as

$$\pi \equiv Y^S - wN^D - vR^D \tag{2.3}$$

where Y^S is the firm's supply of output goods, N^D its demand for labour services, R^D its demand for resources, and where $w \equiv W/P$ and $v \equiv V/P$ are exogenous to the firm, that is, independent of the firm's choice of Y^S, N^D, and R^D.

The firm chooses Y^S, N^D, and R^D to maximize its profits. The short-run production function, which relates the quantities of inputs of labour services and resource flows to the quantity of output, constrains the set of feasible choices. For the representative firm, the short-run production function is assumed to be

$$(2.4) \qquad Y = F(N, R)$$

that is,

$$(2.5) \qquad Y^S = F(N^D, R^D)$$

where F(·) exhibits a positive and diminishing marginal product with respect to each input, as well as diminishing returns to scale, which occurs because in the short run the amount of capital stock is held constant. It is also assumed that labour services and resource flow are technical complements in production: increasing the flow of one increases the marginal product of the other. Hence the signs of the partial derivatives of F(·) are

$$(2.6) \qquad F_N > 0, \quad R_R > 0, \quad F_{NN} < 0, \quad F_{RR} < 0, \quad R_{NR} > 0.$$

It is assumed that there is no inventory accumulation, although this formulation might be relaxed in a future formulation. Further, it is assumed that costs of quantity adjustments are zero, which excludes the possibility of levels of output or inputs independent of prices or sales: firms set output to be equal to sales at all times and minimize the costs of the factor inputs given this level of output.

First-order conditions for profit-maximization require that the marginal physical product of each input be set equal to its real price:

$$(2.7) \qquad w = F_N(N^D, R^D)$$

$$v = F_R(N^D, R^D).$$

Second-order conditions are that the production function be strictly concave in the neighbourhood of the optimal point in the 3D-space of (Y, N, R), which is equivalent to the Hessian matrix of the production function

$$(2.8) \qquad H \equiv \begin{bmatrix} F_{NN} & F_{NR} \\ F_{NR} & F_{RR} \end{bmatrix}$$

being negative definite at the optimal point. H is negative definite if all characteristic roots are negative, that is, if

$$(2.9) \qquad F_{NN} < 0$$

$$F_{RR} < 0$$

and

$$(2.10) \qquad D_1 \equiv |H| = F_{NN}\,F_{RR} - (F_{NR})^2 > 0.$$

It is assumed that $(F_{NR})^2 < F_{NN}\,F_{RR}$ so that H is negative definite throughout. Then the first-order conditions are necessary and sufficient for profit-maximization.

Solution of the first-order conditions results in expressions for the notional demand for labour services, the notional demand for resource flow, and notional supply of output, all in terms of the real wage rate and the real resource flow price:

$$(2.11) \qquad N^D = N^D(w, v)$$

$$R^D = R^D(w, v)$$

$$Y^S = F(N^D, R^D) = Y^S(w, v).$$

That is, the short-run notional supply of output is the output ($Y = Y^S$) corresponding to the employment ($N = N^D$) at which the marginal product of labour (F_N) equals the real wage and to the resource use ($R = R^D$) at which the marginal product of resource flow (F_R) equals the price of resource flow in real terms.

Standard analysis of Appendix A1 yields the comparative statics equations:

$$\text{(2.12)} \qquad \begin{aligned} N_w^D &= F_{RR}/D_1 < 0 \\ N_v^D &= -F_{NR}/D_1 < 0 \\ R_w^D &= -F_{NR}/D_1 < 0 \\ R_v^D &= F_{NN}/D_1 < 0 \\ Y_w^S &= (wF_{RR} - vF_{NR})/D_1 < 0 \\ Y_v^S &= (vF_{NN} - wF_{NR})/D_1 < 0 \end{aligned}$$

where

$$D_1 \equiv F_{NN}\,F_{RR} - (F_{NR})^2 > 0.$$

Thus profit maximization subject to the assumed production function implies that notional labour demand, notional resource flow demand, and notional output supply are all inversely related both the the real wage and to the real resource flow price.

2.2.2. The households.

The households are modelled by the standard treatment of household behaviour. The households are assumed to receive the profits, the wages, and the resource receipts, less income taxes (T), which in total is the disposable income, Y^{DI}

$$(2.13) \qquad Y^{DI} \equiv wN^S + vR^S + \pi - T$$

and when there is market-clearing in all three markets this becomes

$$(2.14) \qquad Y^{DI} = Y - T,$$

that is, the disposable income of the households equals the value in real terms of the actual sales less the total income tax.

The households buy consumables, and save from their income. Saving and asset accumulation, unlike consumption, are generally undertaken not for their own sake, but for the sake of future consumption, and thus it is rather difficult to fit the saving decision into the conventional static theory of allocation of household income among expenditures on different goods. In order to do this three simplifications of full dynamic analysis are made. First, all expenditure on consumer goods is aggregated so that consumption expenditure represents demand for a single aggregated consumer good. Second, all individual decisions on saving are aggregated. Third, this aggregate saving flow is described by a very simple rule relating savings to income.

Savings are assumed to be a positive and increasing function of disposable income. Consumption can be obtained as the residual after savings is subtracted from disposable income, and hence consumption will also generally be a function of disposable income, but this treatment is not to assume that households actually make their decisions of how to dispose of their income in this manner.

Thus

$$Y^{DI} = C^D(Y^{DI}) + S(Y^{DI})$$

where C^D is demand for consumption, and S is total savings as flows. The assumption is made that a very simple rule relates savings to income: that the average and marginal propensities to save out of disposable income are equal and equal to s. Then the expression for consumption demand can be written

$$C^D(Y^{DI}) = (1-s)Y^{DI}. \tag{2.15}$$

Households supply labour services. In more general models the quantity of labour supplied might be modelled as a non-decreasing function of the real wage, but it suits our purposes in this model to assume inelastic labour supply:

$$N^S = \bar{N}^S,\ \bar{N}^S_w = 0. \tag{2.16}$$

2.2.3. <u>The resource suppliers</u>.

As mentioned above, the assumption that the resource suppliers attempt to maximize the returns to their portfolios of assets including stocks of resources leads to the net supply of resource flow's being a decreasing function of the anticipated proportional rate of change of resource price. Ceteris paribus, the supply of resource flow is thus inelastic with respect to the spot price of the resource. Only if the anticipated proportional rate of price change is elastic with respect to the spot price will the supply of resource flow change as the spot price changes. This possibility is examined below when various modes of expectation formation are examined.

The expected, or anticipated, proportional rate of change of the real price of the resource is designated as e, where

$$(2.17) \qquad e(t) \equiv \dot{v}(t)^{e}/v(t);$$

that is, e is defined as being the expected rate of change of real resource price at any time t divided by the spot price at time t. For the moment we assume that the expectations of the resource suppliers are static, despite their experience of changing actual rates of change of price. Further, we assume that given this constant expectation

$$(2.18) \qquad e = \bar{e},$$

there is a positive flow of resource for sale as a factor input to production

$$(2.19) \qquad R^{S} = R^{S}(\bar{e}) \equiv \bar{R}^{S} > 0$$

with

$$\bar{R}^{S}_{v} = 0.$$

2.2.4 The government.

In the basic model, the government can control G, the flow of government purchases of goods and service, and T, the flow of tax revenues (net of transfer) from the households. The government's behaviour is treated as exogenous, and with it the level of the two variables G and T.

2.2.5. The investors.

The basic model includes I, the flow of purchase of output as investment goods. This variable is treated as exogenous.

2.3. Comparative statics.

2.3.1. Market-clearing conditions.

The basic model assumes that exchange takes place only under market-clearing conditions. These conditions require a harmonization of the behaviour of firms, households, and resource suppliers as examined below. In the basic model, exchange takes place in three markets: the labour market, the resource flow market, and the market for homogeneous output.

In the labour market, market clearing requires that both the quantity of labour services demanded by the firms and the quantity of labour services supplied by the households be equal to the actual level of employment. Thus the labour-market-clearing condition is

$$(2.20) \qquad N^D(w, v) \;=\; \bar{N}^S \;=\; N.$$

In the resource flow market, market clearing requires that both the quantity of resource flow demanded by the firms and the quantity of resource flow supplied by the resource suppliers be equal to the actual level of resource flow use. Thus the market-clearing condition for the resource flow market is

$$(2.21) \qquad R^D(w, v) \;=\; \bar{R}^S \;=\; R.$$

In the market for homogeneous output, market clearing requires that both the quantity of output supplied by the firms and the aggregate demand for output be equal to actual production. Thus the market-clearing condition for the market for output is

$$(2.22) \qquad Y^S(w, v) \;=\; Y^D \;=\; Y.$$

Thus market clearing leads to three equations in the two unknowns w and v. If the three are consistent, then any one can be obtained from the other two, and any two of the three are sufficient for solution of the two real prices. In economic terms, the condition of clearing in any two of the markets will fully determine the condition of the third, which will clear. But is there any reason to expect consistency in the system of three equations: given the exogenous levels of supply of labour services and resource flow, is there any reason to expect that clearing in the markets for labour and resource flow will coincide with clearing in the market for homogeneous output? Is there any reason to expect that supply of output at full employment (of both labour services and resource flow) will just equal aggregate demand for output? Examination of the aggregate consumption function will show that for this model there is no reason to expect consistency in the system of three equations, that is, given exogenous technology and exogenous levels of demand for investment goods, together with exogenous supply of the two factor inputs, arbitrary levels of government purchases of goods and services and of taxation will in general lead to an economy in which clearing in any two markets will not coincide with clearing in the third. That is, with exogenous government and investment sectors and no government budget constraint, Walras' Law will not in general hold. Thus the model suggests the traditional rationale for fiscal policy to vary government expenditure and taxation so as to equate aggregate demand for output with full employment output.

Aggregate demand is the sum of the demands for consumption goods (private and government) and for investment goods. Hence

$$(2.23) \qquad Y^D \equiv C^D + I + G$$

where the two exogenous variables I and G are not superscripted since the demands for investment goods and government goods and services are assumed always to be met with supply equalling demand equalling actual amount transacted.

From equation (2.15) above the demand for consumables is

$$C^D = (1 - s)Y^{DI}$$

and the *ex ante* disposable income is given by equation (2.13)

$$Y^{DI} = w\bar{N}^S + v\bar{R}^S + \pi - T$$

where the *ex ante* profits are given by equation (2.3)

$$\pi = Y^S - wN^D - vR^D.$$

Thus, from equation (2.23),

$$(2.24) \qquad Y^D = (1-s)(Y^S - T + w(\bar{N}^S - N^D) + v(\bar{R}^S - R^D)) + I + G$$

which is the *ex ante* disequilibrium consumption function.

When market clearing occurs in the two factor input markets the disposable income is given by

$$(2.25) \qquad Y^{DI} = Y^S - T$$

and the level of aggregate demand is given by

$$(2.26) \qquad Y^D = (1-s)(Y^S - T) + I + G.$$

The market for output clears when

$$Y^D = Y^S = Y,$$

which can occur if and only if

$$(2.27) \qquad Y = \bar{Y} \equiv (I + G - (1-s)T)/s.$$

But clearing in the factor input markets leads to full-employment output

$$(2.28) \qquad Y^S = F(\bar{N}^S, \bar{R}^S).$$

In general, there is no reason to expect that full-employment output will equal $\bar{Y}$. This provides a rationale for government fiscal actions to vary exogenous demand $\bar{Y}$ through changes in taxation or government spending to set aggregate demand equal to full-employment output.

This can be shown on a standard Keynesian cross diagram, Figure 2.1. We see that demand and supply for output are equal only where both equal $\bar{Y}$. If $Y^S < \bar{Y}$, then $Y^S < Y^D < \bar{Y}$, the case of excess demand for output. If $Y^S > \bar{Y}$, then $Y^S > Y^D > \bar{Y}$, the case of excess supply of output.

Thus simultaneous clearing of the three markets can occur if and only if

$$Y = F(\bar{N}^S, \bar{R}^S)$$

and

$$Y = (I + G - (1-s)T)/s.$$

The second of these equations can be rewritten as

$$(Y-T)s - I + T - G = 0,$$

that is,

$$(2.29) \qquad (S-I) + (T-G) = 0,$$

where S is the total flow of savings. This is the standard formulation

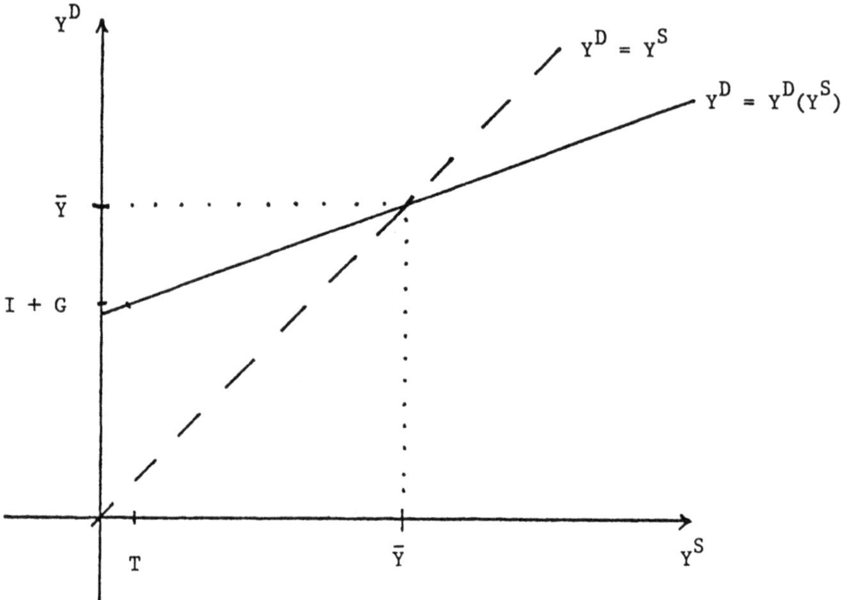

$$Y^D = (1-s)(Y^S - T) + I + G$$

$$\bar{Y} \equiv (I + G - (1-s)T)/s$$

Figure 2.1 The standard Keynesian cross

which has been used to justify government fiscal action to stimulate demand in times of low investment.

Note that if the government balances its budget (i.e. $T = G$), then the level of aggregate demand is

$$(2.30) \qquad Y^D = (1-s)Y^S + sT + I,$$

which shows that using a balanced budget, in the simple model, the government can increase the aggregate level of demand by increasing both government spending and the total tax revenues. For exogenous I and G, and given propensity to save from disposable income, the market for output is cleared when

$$(2.31) \qquad Y^S = Y^D = \frac{I}{s} + T$$

where $T = G$, since the government budget is balanced. Thus if aggregate demand is below the notional supply of output associated with market-clearing in the markets for labour and natural resources, the government, while balancing its budget, can increase the aggregate demand for output by increasing its spending and taxation, as argued in the standard texts.

We assume in this chapter that

$$(2.32) \qquad \bar{Y} = F(\bar{N}^S, \bar{R}^S)$$

so that the three market-clearing equations

$$(2.33) \qquad \begin{aligned} N^D(w, v) &= \bar{N}^S = N \\ R^D(w, v) &= \bar{R}^S = R \\ Y^D(Y^S) &= Y^S(w, v) = \bar{Y} = Y \end{aligned}$$

are consistent. Note that $\bar{Y}$, being exogenous, is inelastic with respect to w and v. The elasticities of the two notional factor input demand functions and the output supply function were derived in section 2.2.1 above as equations (2.12). All are negative.

We can examine the implications of the three equations (2.33) describing market-clearing in the three markets by means of graphical representations. Figure 2.2 represents the labour market. The ordinate is the real wage and the abscissa the quantity of labour services. Notional labour demand depends on w and v, and is represented as a family of downwards sloping curves. The figure shows three of these curves, corresponding to three values of v, where $v_1 < v^* < v_2$. Labour supply is inelastic, and is shown as a vertical line.

It can be seen that clearing in the labour market is consistent with various combinations of w and v, including (w^*, v^*). If v is equal to v^*, then at real wage rates above w^* there would be excess supply of labour services, with $N^D < \bar{N}^S$. At real wage rates below w^* there would be excess demand for labour services. Similarly, with w equal to w^*, real resource price above v^* would lead to excess supply of labour, while real resource price below v^* would lead to excess demand for labour.

Figure 2.3 represents the resource flow market in the resource flow/real resource price plane. Notional resource demand depends on both v and w, and is shown as a family of downwards sloping curves. The figure shows three of these curves, corresponding to three values of w, where $w_1 < w^* < w_2$. Resource flow supply is inelastic, and is shown as a vertical line.

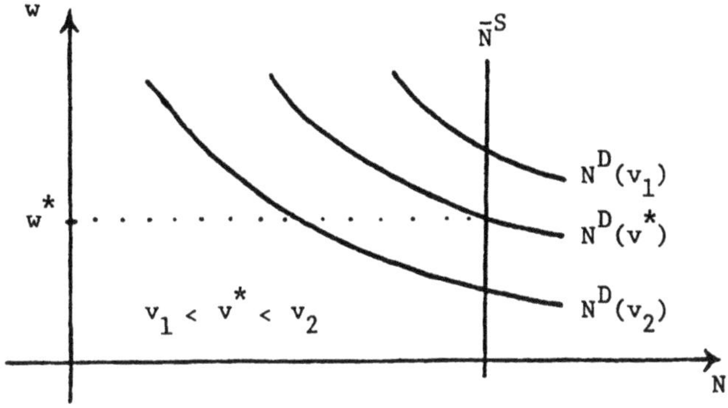

Figure 2.2: The labour market in the basic model.

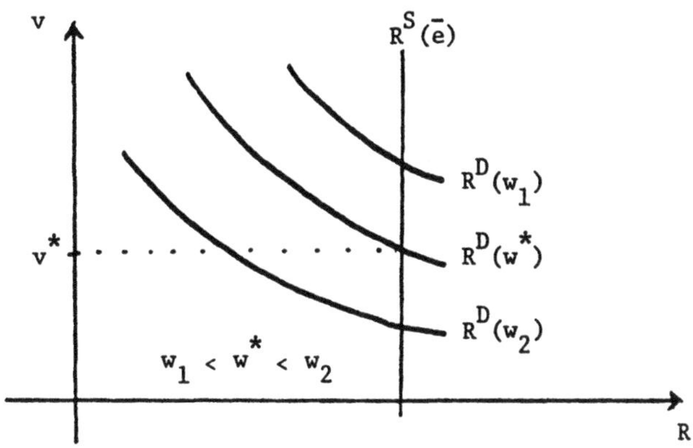

Figure 2.3: The resource flow market in the basic model.

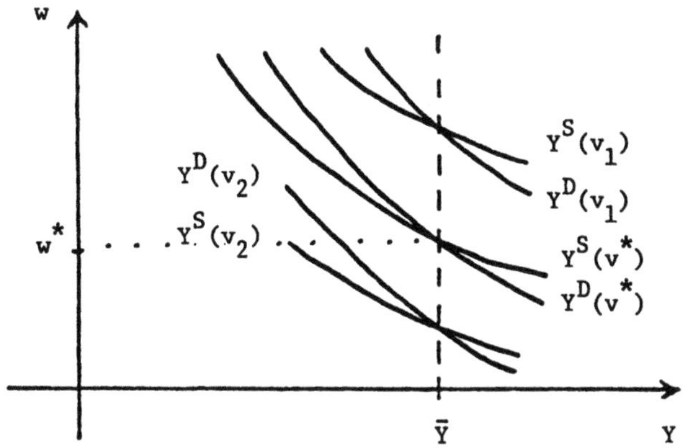

Figure 2.4: The output market in the basic model with $\bar{Y} = F(\bar{N}^S, \bar{R}^S)$.

It can be seen that clearing in the resource flow market is consistent with various combinations of w and v, including (w*, v*). With w equal to w*, v greater than v* would lead to R^D less than $\bar{R}^S$, while v less than v* would lead to R^D greater than $\bar{R}^S$. Similarly, with v equal to v*, w greater than w* would lead to R^D less than $\bar{R}^S$, while w less than w* would lead to R^D greater than $\bar{R}^S$.

Figure 2.4 represents the output market, drawn with

$$\bar{Y} = F(\bar{N}^S, \bar{R}^S).$$

The ordinate is the real wage axis, and the abscissa the quantity of output axis. Notional output supply depends on both w and v, and can be represented as a family of downwards sloping curves. The figure shows three of these curves, corresponding to three values of v, where $v_1 < v^* < v_2$. Notional aggregate demand is a decreasing function of notional output, as seen in Figure 2.1. Hence in Figure 2.4 the three aggregate demand curves are drawn more steeply than the notional output curves. Each pair of curves, corresponding to a particular value of v, is seen to intersect at $\bar{Y}$, which is assumed to equal the full employment output.

It can be seen that clearing in the output market, as in the other two, is consistent with various combinations of w and v, including (w*, v*). With v equal to v*, w greater than w* would lead to Y^D greater than Y^S, while w less than w* would lead to Y^D less than Y^S. Similarly, with w equal to w*, v greater than v* would lead to Y^D greater than Y^S, while v less than v* would lead to Y^D less than Y^S.

The combination (w^*, v^*) is consistent with clearing in both markets. It can easily be shown that this combination is unique. To do so, we combine the analysis of Figures 2.2, 2.3, and 2.4 into a single diagram, Figure 2.5. In this figure, $N^D = \bar{N}^S$ represents the locus of values of w and v consistent with clearing in the labour market. The negative partial derivatives of $N^D(w, v)$ imply that this locus is downwards sloping: an increase in w will have to be accompanied by a reduction in v to maintain a constant demand $N^D(w, v) = \bar{N}^S$. The inequality signs on each side of the labour market clearing locus are to indicate that for combinations of w and v to the NE of the locus there is excess supply of labour ($N^D < \bar{N}^S$), and that for combinations to the SW there is excess demand ($N^D > \bar{N}^S$).

The line labelled $R^D = \bar{R}^S$ represents the locus of values of w and v consistent with clearing in the resource flow market. The negative partial derivatives of $R^D(w, v)$ imply that this locus too is downwards sloping. For combinations of w and v to the NE of the locus there is excess supply of resource flow ($R^D < \bar{R}^S$), and to the SW there is excess demand ($R^D > \bar{R}^S$).

The line labelled $Y^D = Y^S$ represents the locus of values of w and v consistent with clearing in the output market. From Figures 2.1 and 2.4 this can only occur when $Y^D = Y^S = \bar{Y}$. The negative partial derivatives of $Y^S(w, v)$ imply that this locus is downwards sloping: an increase in w will have to be accompanied by a reduction in v to maintain a constant supply $Y^S(w, v) = Y^D(Y^S) = \bar{Y}$. The market-clearing locus separates the excess demand region ($Y^D > Y^D$) to the NE from the excess supply region ($Y^D < Y^S$) to the SW.

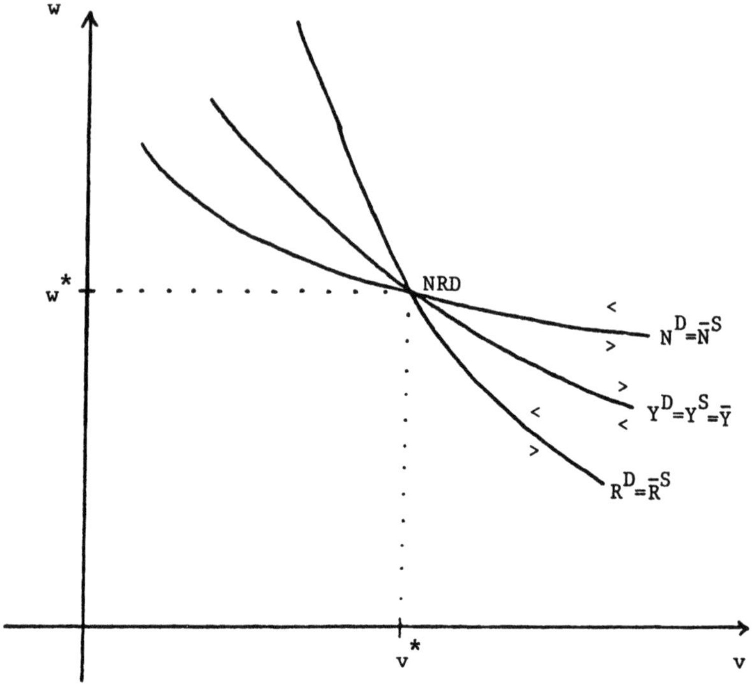

Figure 2.5: Notional market-clearing loci with $\bar{Y} = F(\bar{N}^S, \bar{R}^S)$.

The relative slopes in Figure 2.5 are shown in Appendix A2 to follow if the production function is Cobb-Douglas with diminishing returns to scale. The resource- and labour-market-clearing loci are as shown if the marginal product of each factor input diminishes more rapidly as the level of that factor increases than it increases with an increase in the other factor, ceteris paribus (inequalities A13). This is a common assumption.

The three market-clearing loci intersect at the point (w*, v*), and the monotonicity of the loci means that this combination of real prices is uniquely consistent with clearing in all three markets. The combination (w*, v*) also implies values for N, R, Y, C, W, V, and P consistent with general market clearing.

It is instructive to examine how Figure 2.5 would appear if the government did not set the exogenous demand $\bar{Y}$ to equal the full-employment output. Figure 2.6a has been plotted for the condition of understimulation:

$$(2.34) \qquad \bar{Y} < F(\bar{N}^S, \bar{R}^S).$$

The output market-clearing locus is to the NE of the point NR, the intersection of the labour and resource market-clearing loci, which is the combination of w and v such that both factor input markets clear. From the discussion above, as summarized in Figure 2.5, to the NE of the output market-clearing line there is excess demand. Moving SW with decreasing w and v, notional output Y^S increases, as shown by the negative partial derivatives in section 2.2.1, as more slowly does notional aggregate demand Y^D, as shown in Figure 2.1, until clearing occurs in the output market. For lower $\bar{Y}$, clearing occurs at lower output, at higher w and v as shown in Figure 2.6a, with the output market-clearing locus to the NE of its position in Figure 2.5. Point RD

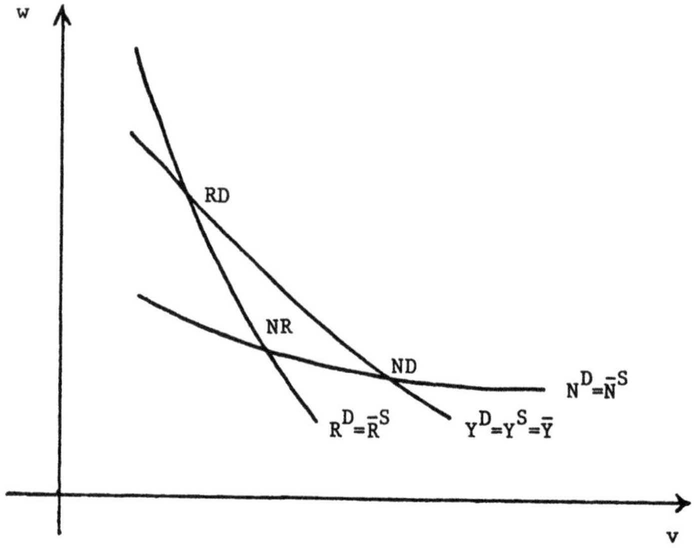

Figure 2.6a: Notional market-clearing loci with $\bar{Y} < F(\bar{N}^S,\bar{R}^S)$.

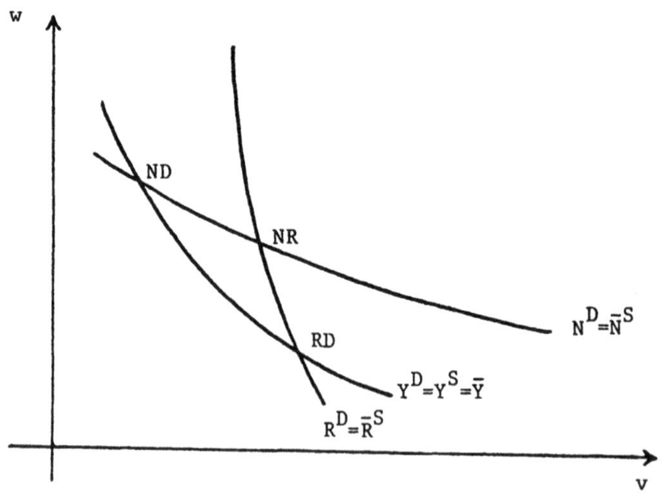

Figure 2.6b: Notional market-clearing with $\bar{Y} > F(\bar{N}^S,\bar{R}^S)$.

is the combination of w and v at which the markets for output and resource flow clear, and point ND the combination at which the markets for output and labour services clear. In the triangular region between RD, ND, and NR there is excess supply on all markets.

Figure 2.6b has been plotted for the condition of overstimulation:

$$(2.35) \qquad \bar{Y} > F(\bar{N}^S, \bar{R}^S).$$

The higher $\bar{Y}$ has meant that output market clearing occurs at higher output Y^S than in Figures 2.5 or 2.6a, at combinations of w and v to the SW of the output market-clearing locus of Figure 2.5. The points RD, ND, and NR designate the same market-clearing conditions as in Figure 2.6a. In the triangular region between them there is excess demand on all markets.

2.3.2. The effects of exogenous disturbances

Analysis of the comparative statics of the basic model is concerned with the relationships between the exogenous variables of the model and the values of the endogenous variables which satisfy the market-clearing conditions. Since the system is over-determined, if only one exogenous variable changes the system may become inconsistent, which will happen if $\bar{Y}$ no longer equals the full-employment output, $F(\bar{N}^S, \bar{R}^S)$. Possible exogenous disturbances include changes in supply of resource flow, in production technology, in the household propensity to save, and in government fiscal policy. We shall assume that any change in one of the former exogenous variables is immediately accompanied by the change in government fiscal policy necessary to ensure that the general market-clearing condition is obeyed:

$$F(\bar{N}^S, \bar{R}^S) = \bar{Y} \equiv (I + G - (1-s)T)/s. \tag{2.36}$$

Given a change in one of the exogenous variables (and, if necessary, the accompanying change in government fiscal policy), what changes, if any, will be required in w^* and v^* in order to satisfy the market-clearing conditions? The process by which the endogenous variables change is the subject of the dynamic analysis below.

An increase in $\bar{R}^S$ will require either an increase in G or a decrease in T or some combination of both to ensure that the above condition is obeyed. Consider the labour market-clearing locus. An increase in $\bar{R}^S$, although it may change w^* and v^*, will not alter the functional dependence of the demand for labour on w and v, nor will it affect the inelastic supply of labour. Hence the labour market-clearing locus is not shifted, and the new combination (w^*, v^*) must lie on it. Since the supply of resource has increased, the resource market will at first be in a state of excess supply. From the discussion above, this implies that the resource market-clearing locus has moved to the SW, leaving the economy at first in the region of excess supply. The new general market-clearing point in Figure 2.7 will be to the NW along the $N^D = \bar{N}^S$ locus, with an increase in w^* from w_1^* to w_2^* and a decrease in v^* from v_1^* to v_2^*. Since government fiscal policy is assumed to follow in response to changes in other exogenous variables, the output market-clearing locus will pass through the new point (w_2^*, v_2^*). For clarity, it has been omitted from Figure 2.7. A decrease in $\bar{R}^S$ would result, after appropriate government fiscal response, in a decrease in w^* and an increase in v^*.

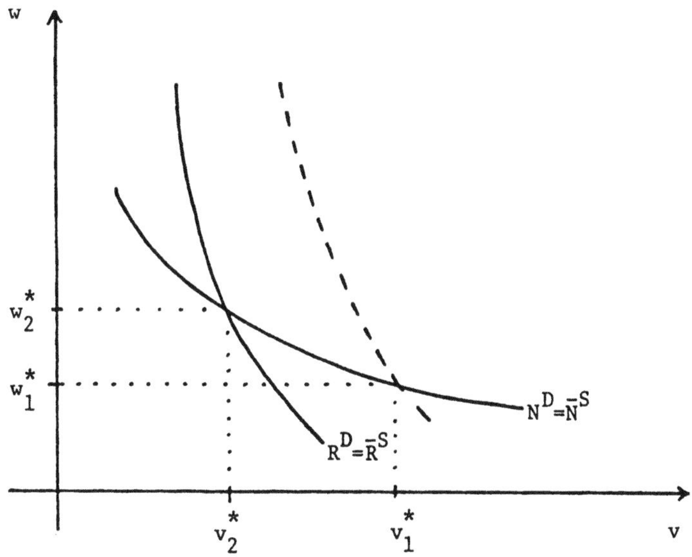

Figure 2.7: Effect of an increase in resource supply, $\bar{R}^S$.

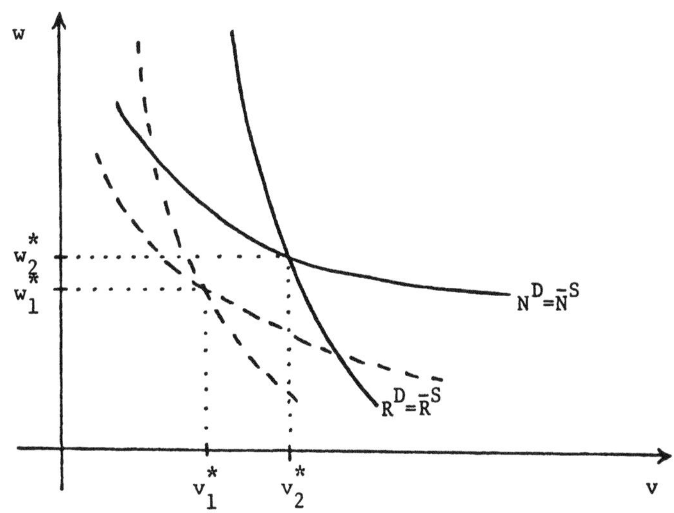

Figure 2.8: Effect of an improvement in production technology.

We shall examine two kinds of improvement in production technology:

(i) technical progress in which a given output Y can be produced by a given labour input N combined with a resource input R smaller than before. Such technical progress can be characterized as "resource-augmenting" and can be formalized as

$$(2.37) \qquad Y = F^1(N, R) \equiv F(N, \alpha R)$$

where $\alpha > 1$. This is similar to Solow-neutral technical progress which is of the same form except that it is capital-augmenting instead of resource-augmenting;

(ii) technical progress in which the production function shifts by a uniform upwards displacement of the whole function. This disembodied form of technical progress is similar to Hicks-neutral technical progress except that the second variable factor input is resource flow instead of capital. It can be formalized as

$$(2.38) \qquad Y = F^2(N, R) \equiv \alpha F(N, R), \qquad \alpha > 1.$$

Consider $F^2(N, R)$. For any given (w, v), examination of the first-order conditions for profit maximization shows that as α increases the two marginal products F_N and F_R must decrease, which means that N^D and R^D must increase, given strictly concave production technology. Thus the effect of increasing α is to shift the N^D and R^D functions, shown in Figures 2.2 and 2.3, to the right. At any real wage and real resource flow price firms will want to use more labour and buy more resource flow. This increase in factor inputs will be converted more efficiently into output, and so firms will want to sell more output. As a result of these shifts, the values of w and v which originally satisfied the market-clearing conditions will no longer do so. Analysis in Appendix A3 shows that w* and v* probably both increase to re-establish the market-clearing conditions. Figure 2.8 illustrates the shift to the right of the resource market-clearing locus and the shift upwards of the labour market-clearing locus. The dashed lines indicate the old loci, and the original market-clearing point (w_1^*, v_1^*) is seen as implying excess demand for both factor inputs and excess supply of output.

Consider $F^1(N, R)$. Analysis in Appendix A3 shows that N^D will always move to the right in Figure 2.2, while for large α, R^D may move to the left in Figure 2.3. Figure 2.8 illustrates technical change for small α with both w* and v* increasing. For large α, such that $1/\alpha$ is less than the resource elasticity of the marginal product of resource, analysis in the appendix shows that w* will rise while v* falls.

A change in the household propensity to save s will not alter the fundamental dependence of the demands for the factor inputs on w and v, nor will it affect the inelastic supplies of factor inputs. Hence neither of the factor input market-clearing loci in Figure 2.5 is shifted by the change, and the market-clearing combination (w*, v*) is unchanged. But without a change in government fiscal policy, the output market will no longer clear: a reduction in the saving propensity will require a reduction in the government deficit (G - T) to maintain equality between $\bar{Y}$ and full-employment output, and an increase in s will require an increase in the deficit.

2.4. Dynamic analysis in the basic model

2.4.1. The basis for notional schedules

We have assumed that all agents are price-takers and that exchange takes place only under market-clearing conditions: the supply and demand schedules are thus "notional" and each agent's *ex ante* assumption that he can buy and sell any amount desired at the existing real prices w and v, with no quantity rationing of buyers or sellers, is always borne out *ex post* by the actual transactions which occur in the markets. If exchange took place without market clearing then some agents, those on the long side, would find themselves rationed. In a market with excess supply sellers would be unable to sell all they had planned to, given the going prices, and in a market with excess demand buyers would be unable to buy all they had planned to, given the going prices. This situation of rationing of the long side in non-market-clearing exchange will be examined further in the following chapters.

But if exchange takes place only under market-clearing conditions, how is the market-clearing combination (w^*, v^*) determined and reached by the economy? The assumption of trading only at market clearing has been explained by the mythical Walrasian auctioneer who announces a price to the possible traders in a market, observes the resulting offers to buy and sell, and announces a new price revised up or down in an attempt to reduce the difference between aggregate demand and supply. If no transactions took place while this "tatonnement" (groping) process was occurring (perhaps because of the process' taking place too quickly) then the price at which aggregate demand and supply were equal could eventually be determined and trading take place at this price. If trading took place before equality between aggregate demand and supply had been reached, the process could still attain a price at which the market cleared, but if any sales had occurred during the process it is unlikely that this market-clearing price would be the price obtained with no trading during the process: those traders who had completed transactions during the process would have left the market, and their offers to buy and sell would no longer appear in the aggregate demand and supply considered by the Walrasian auctioneer.

To allow the *ex ante* market-clearing combination (w*, v*) obtained from the notional schedules to equal the *ex post* combination resulting from the tatonnement process, the assumption is made either that no trading occurs during the process or that such transactions as do occur are recontractable after the market-clearing combination is obtained. These unrealistic assumptions are relaxed in the following chapters. But even with them, convergence is not assured in our three-market model. If we postulate one Walrasian auctioneer for each market the process may take many iterations since the firm's offers to buy or sell on one market are not only a function of the (nominal) price on that market but also a function of the prices on the other two markets, which may be changing independently. Only when there is no tendency for change on any market (equality of supply and demand on each market) will the combination (w*, v*) have been obtained.

2.4.2. Price adjustments

For the moment we continue the assumptions of a price-setting Walrasian auctioneer for each market, and no trading until the tatonnement process has converged to the market-clearing combination (w*, v*). Alternative formulations will be discussed at greater length in a following chapter but for the moment we assume that the activities of these mythical auctioneers in revising the nominal price on each market to reduce the difference betweeen aggregate demand and supply on that market can be modelled by the Walrasian excess demand hypothesis, which in general form can be written

$$(2.39) \qquad \frac{1}{p}\frac{dp}{dt} = f(q^D - q^S) = f(q^X), \qquad f' > 0$$

where dp/dt is the time rate of change of nominal price, and q^X is the notional excess demand on the market. With no autonomous price increases f(0) is zero.

For the labour market the hypothesis can be written

$$(2.40) \qquad \dot{W}/W = \lambda_W N^X$$

where the positive constant λ_W is the speed of money wage adjustment. This equation states that the proportional rate of change of the nominal wage is an increasing function of the notional excess demand on the labour market.

For the resource flow market the hypothesis can be written

$$(2.41) \qquad \dot{V}/V = \lambda_V R^X$$

where the positive constant λ_V is the speed of adjustment of the money price of resource flow. The equation states that the proportional rate of change of the nominal price of resource flow is an increasing function of the notional excess demand on the market for resource flow.

For the output market the hypothesis can be written

$$(2.42) \qquad \dot{P}/P = \lambda_P Y^X$$

where the positive constant λ_P is the speed of adjustment of the money price of output, a proxy for the overall price level. This formulation models "demand-pull" price inflation, with no autonomous or "cost-push" component, and states that the proportional rate of change of the nominal price of output is an increasing function of the notional excess demand on the market for output.

The proportional rates of change of the real wage and the real price of resource flow are given by

$$(2.43) \qquad h \equiv \dot{w}/w = \dot{W}/W - \dot{P}/P = \lambda_W N^X - \lambda_P Y^X$$

$$g \equiv \dot{v}/v = \dot{V}/V - \dot{P}/P = \lambda_V R^X - \lambda_P Y^X.$$

Since the notional excess demands N^X, R^X, and Y^X, as derived above, are functions of w and v, it is possible to plot a phase diagram for the economy, showing the tendency for the two real prices to change in the different areas of the plane.

Consider Figure 2.9, which is similar to Figure 2.5 and shows the three market-clearing loci. The positive quadrant has been divided into six regions, and in each region the signs of the three rates of change of nominal price are shown as a vector:

$$(\dot{P}/P,\ \dot{W}/W,\ \dot{V}/V).$$

Figures 2.10a and 2.10b, similar to Figures 2.6a and 2.6b respectively,

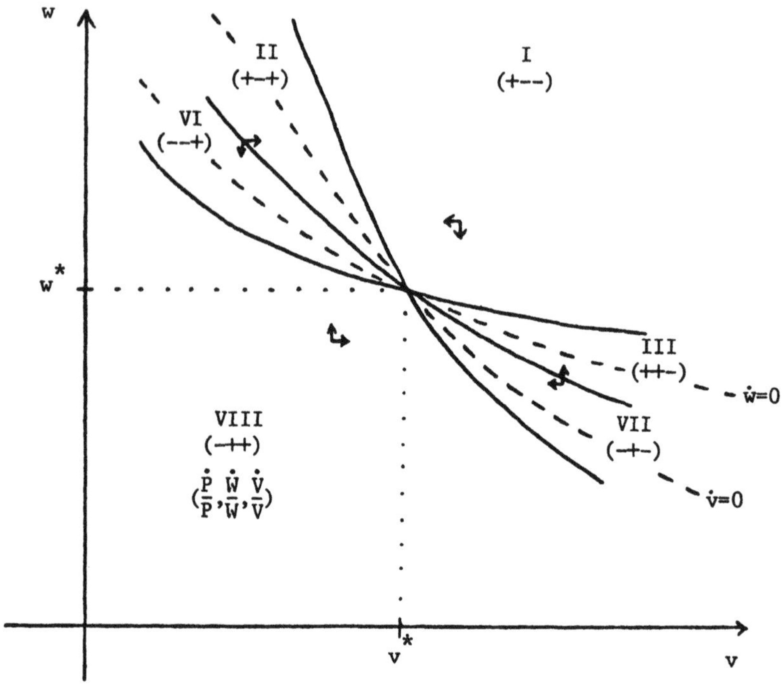

Figure 2.9: Real price dynamics in the basic model, $\bar{Y} = F(\bar{N}^S, \bar{R}^S)$.

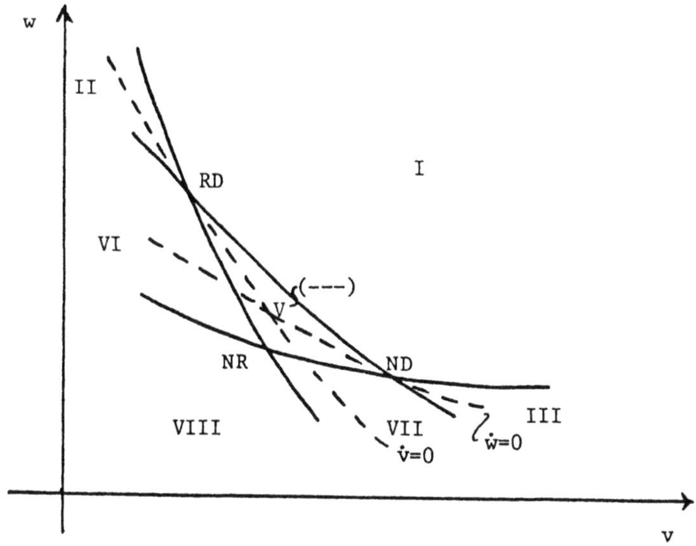

Figure 2.10a: Real price dynamics in the basic model, $\bar{Y} < F(\bar{N}^S,\bar{R}^S)$.

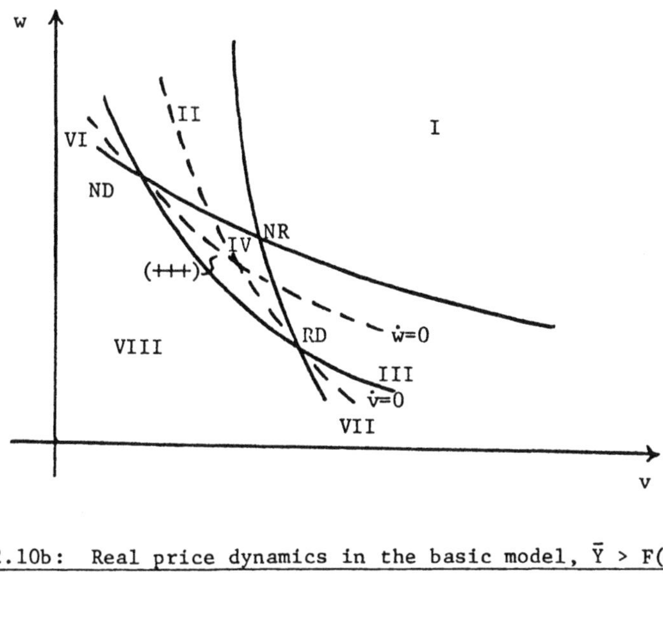

Figure 2.10b: Real price dynamics in the basic model, $\bar{Y} > F(\bar{N}^S,\bar{R}^S)$.

include the two additional regions which occur when the autonomous demand $\bar{Y}$ does not equal the full-employment output. The eight regions are summarized in Table 2.1, which also includes the signs of the rates of change of the two real prices in each region.

In plotting a phase diagram we are concerned with the loci of constant-w and constant-v. In regions II, IV, V, and VII, $\dot{P}/P$ is the same sign as $\dot{V}/V$ so that $\dot{v}/v$ will be positive or negative depending on the relative sizes of the two. When they are equal, $\dot{v}/v$ is zero

$$g \equiv \dot{v}/v = \dot{V}/V - \dot{P}/P = 0. \tag{2.44}$$

The constant-v locus passes through regions II, IV, V, and VII, and through the points RD and NRD where the resource flow and output markets clear. Similarly, we can show that the constant-w locus passes through regions III, IV, V, and VI, and through the points ND and NRD where the labour and output markets clear. The loci are shown in Figures 2.9 and 2.10. To the left of the constant-v locus, $\dot{v}$ is positive, and to the right, negative. Below the constant-w locus, $\dot{w}$ is positive, and above, negative. The small arrows in the figures indicate these movements.

2.4.3. The dynamics of adjustment

We should like to be able to describe the time paths followed as the basic model adjusts to a change in one of the exogenous variables as considered in section 2.3.2 above. For the economy of Figure 2.9 we can use the price adjustment equations of section 2.4.2 to derive the time path as long as we remember that we have assumed either that no trading occurs until the general market-clearing point NRD is reached,

Table 2.1. The eight regions with notional schedules.

Region	$\dot{P}/P,\ Y^X$	$\dot{W}/W,\ N^X$	$\dot{V}/V,\ R^X$	$h \equiv \dot{w}/w$	$g \equiv \dot{v}/v$
I	+	-	-	-	-
II	+	-	+	-	±
III	+	+	-	±	-
IV	+	+	+	±	±
V	-	-	-	±	±
VI	-	-	+	±	+
VII	-	+	-	+	±
VIII	-	+	+	+	+

or that such trading as does occur before then is recontractable after the general market-clearing combination is revealed.

Figure 2.11 shows adjustment paths to the point NRD from various initial combinations of (w, v). Analysis in Appendix A4 shows that the general market-clearing equilibrium point NRD is locally dynamically stable if the propensity to save s is small or if the speed of adjustment of the money price of output λ_p is small.

But the economies of Figures 2.10a and 2.10b do not have a general market-clearing point. There do exist three points at which pairs of markets clear, but the third market in each case does not clear. In such cases, although we have derived points at which the two state variables w and v are constant (the intersections of the constant-w and constant-v loci), trading will not take place under the assumption of no trading until the general market-clearing price combination has been attained, since this combination does not exist if

$$\bar{Y} \neq F(\bar{N}^S, \bar{R}^S).$$

Similarly, under the alternative assumption of recontracting, exchange may never take place.

But it is quite likely that government, even if it wanted to do so, could not adjust the autonomous demand so as to set $\bar{Y}$ equal to the full-employment output: it may be uncertain just what full-employment output is, or what the level of exogenous demand for investment goods is, or it may find that such a fiscal policy is inconsistent with other goals of economic management. For this reason the basic, market-clearing model is unsatisfactory. The model is unsatisfactory too in its

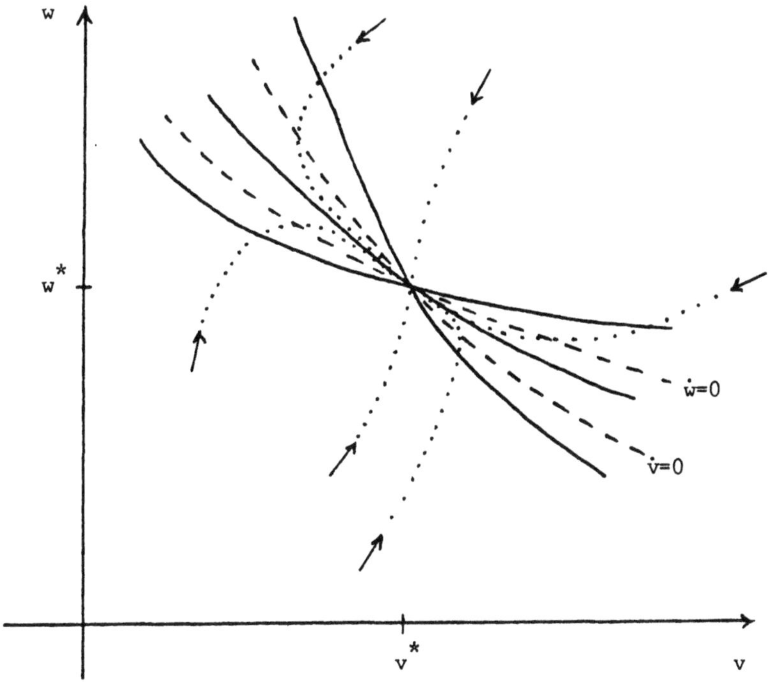

Figure 2.11: Convergence to market-clearing conditions.

assumptions of recontracting or no trading before market clearing: in reality there is trading, with binding contracts. The next chapter will present a model in which disequilibrium trading can occur, and in which the government is not assumed to set $\bar{Y}$ equal to full-employment output.

CHAPTER III: A SIMPLE FIX-PRICE DISEQUILIBRIUM MODEL

3.1. Non-market-clearing exchange.

The basic model of above determined the quantities of the economic goods and their exchange ratios by analyzing the relationships between the exogenous variables of the model and the values of endogenous variables which satisfy the market-clearing conditions. As discussed, market clearing can be achieved with the assumption of Edgeworth's recontracting or Walras' tatonnement, in which no trading occurs at "false" prices.

But recontracting does not characterize actual markets. In reality, offers to buy and sell are usually binding, and most actual exchanges occur at "false," that is, non-market-clearing, prices. The analysis below does not assume trading at market clearing only. It examines the determination of output, employment, and resource flow under non-market-clearing conditions. In analysis of the disequilibrium model, the starting point is some momentary situation in which all the variables of the system have assumed given values. If the momentary situation is not an "equilibrium," certain variables adjust so that in the short run an "equilibrium" (possibly a quasi-equilibrium) is reached. In this model monetary factors are abstracted from: it is assumed that the monetary authorities manage to keep the rate of interest constant, and that monetary variables do not influence aggregate demand. In the short run the stock of capital goods, the labor supply function, and the stock of resource are assumed constant.

Dropping the assumption of trading only at market clearing has two essential implications for the determination of the quantities traded. First, the quantities traded cannot be determined simply by reference to market-clearing conditions. There is no equivalence between actual transactions and quantities supplied and demanded. In order to analyze quantity determination under non-market-clearing conditions, the actual trading process must be examined. The assumption of "voluntary exchange" will be made: no economic actor can be forced to buy more than he demands or to sell more than he supplies. Consequently, the actual level of total transactions will be determined by the "short" side of the market (that is, by suppliers if there is excess demand, by demanders if excess supply), and economic actors on the "long" side will be constrained in their transactions (which implies some rationing scheme).

These constraints lead to the second implication: not every economic actor will generally act as if he can sell or buy any amount which he supplies or demands at the existing price vector. In particular, economic actors on the "long" side of the market (that is, suppliers if there is excess supply, demanders if there is excess demand) will face quantity constraints on their transactions, to be taken into account when formulating behaviour in other markets. These modified demand functions are called "effective" demand functions. The notional (or unconstrained) demand and supply functions derived in the basic market-clearing model above do not in general describe the behaviour of the firms and households.

The existence of excess supply or demand in one market implies the likely divergence of the effective supply or demand from the notional supply or demand in each of the other two. The economic actor must derive his supply and demand functions taking into account fully the information about other markets.

The present chapter examines the determination of effective supplies and demands and analyzes their implications for the determination of output, employment, and resource flow under non-market-clearing conditions. For the beginning, the nominal prices of output, labor, and resource are taken as given, and the levels of output, employment, and resource use implied by the particular price vector are calculated. This represents an extreme case, which implies that the momentary level of each of the quantities depends on the price vector of that moment, which suggests the possible existence of an ultra-short-run quantity-adjustment process occurring within the "moment" examined here: a moment long enough "for the multiplier to work itself out" (Leijonhufvud (1968)) and the quantities to adjust to the target levels implied by the price vector. Ignoring the adjustment of the quantities to their target levels is equivalent to assuming that quantities adjust infinitely faster than prices. Solow and Stiglitz (1968) described a model in which quantity adjustments occur at speeds comparable to the speeds of price adjustment.

3.1.1. Price rigidities in the short run.

A basic assumption in the following model is that the economy can be specified completely at any instant in terms of its prices, the combination of real wage w and real resource flow price v. That is, w and v are the two state variables of the economic system, from which can be deduced all other endogenous variables, including the quantities of output, labour services, and resource flow traded at any (w, v). This is similar to the basic model of the previous chapter, but that model was essentially static, with the explicit assumption of no trading until the Walrasian auctioneer had finished his work of determining the price vector (w, v) at which all three markets would clear (or with the possibility of recontracting after the tatonnement). In relaxing this assumption of no non-market-clearing exchange, we must face the dynamic problem of what determines the quantity traded before market clearing.

Traditional theory teaches that prices react quickly to excess supplies or demands, but recent writers (Leijonhufvud (1968), Alchian (1970), Malinvaud (1977)) have questioned the adequacy of this theory in describing the short-run behaviour of modern economics. In particular they point to better knowledge of how prices are determined and to recent microeconomic analyses of individual decisions to bolster their view that an excess of demand (or supply) leads not to a rapid increase (decrease) in price to reduce the excess but to changes in order-books, queues, inventories, delivery dates, and so on, which have the effect that the discrepancy between demand and supply remains, with the long side of the market (buyers in the case of excess demand) constrained or rationed. That is, short-run quantity adjustments are much more apparent and influential than short-run price adjustments.

Microeconomic analyses of individual behaviour including the relevant uncertainties and transaction or information costs have suggested that quantities will adjust before prices. Okun (1975) spoke of traditional "auction" markets such as markets for agricultural products, raw materials, and most assets, in which prices quickly adjust to equate demand and supply, contrasting them to "customer" markets for services and manufactured goods, in which the existence of long-run contracts means that prices do not react to excess demand or supply: excess demand may mean higher prices in any new contracts made, but another response may be an increase in the number of new contracts--at any rate, the prices in existing contracts would remain unchanged.

It may be expected that markets for stocks of natural resource would be closer to traditional auction markets, but we consider the market for flow, where the resource is bought as a factor input to production, and there is a case for modelling this as an Okun customer market, with firms preferring to ensure medium-run supply of resource by signing contracts at prices perhaps higher than the spot price for resource stocks. This hypothesis could be tested by empirical study. At any rate, the following model includes the assumption that in response to excess demand (or supply) on any of the three markets for output, labour services, or resource flow the quantity adjusts much more rapidly than the price, so that it is not necessary to examine quantity adjustment explicitly: the actual quantity demanded or supplied at any time is assumed equal to the target quantity demanded or supplied, a function of the vector of going prices. For this chapter the prices will be assumed not to adjust: we can think of the speed of adjustment of prices in response to excess demand or supply as being imperceptible in the period under analysis. The analysis will resemble that of the "fix price" method of Hicks (1965).

3.1.2. Properties of a non-market-clearing equilibrium.

In analysis of a period long enough for quantities to adjust to their target values, but not so long that prices begin to adjust in response to excess demand or supply, we need to build a model in which there is consistency between individual actions. With constraints on purchases or sales because of lack of market clearing leading to rationing of the long side of any market, we need to distinguish each individual agent's demand or supply from his purchases or sales. When there is market clearing there is no distinction, with demand equals purchases equals supply equals sales. But without market clearing, purchases (or sales) is the amount actually traded, while demand (or supply) is the amount that the individual agent would like to trade, given the going prices.

How are the target values characterised in the case of non-market-clearing exchange? Malinvaud (1977) argues that there are three general properties necessary for an equilibrium in which, for the given prices, quantities have no further tendency to move. First, trades balance: for each good the sum of purchases equals the sum of sales. Second, there is no involuntary exchange: there is no individual on any market whose purchases exceeds his demand or whose sales exceeds his supply. This property follows from the fact that in a free market no one is forced to trade against his will.

Given the second property, an individual in a market can be in one of four exclusive situations: he can be a constrained or rationed buyer if his (positive) demand exceeds his purchases; he can be an unrationed buyer if his demand equals his purchases; he can be a constrained or rationed seller if his (positive) supply exceeds his sales; he can be an unrationed seller if his supply equals his sales.

The third property deals with consistency in any market. There cannot exist both a rationed buyer and a rationed seller in the same market, for, if so, each would be able to make an advantageous trade. In effect, this property states that there is one and only one market for each good, and that all individual agents have free access to this market.

Given these three properties the target amount traded on any market must be the lesser of the amounts supplied and demanded, which can be written

$$X = \min(X^D, X^S) \tag{3.1}$$

where X is the amount of the good traded, and X^D and X^S are the amounts demanded and supplied on any market. The market for any good is then in one of three states: it can be "balanced," with clearing and no rationing,

$$X = X^D = X^S, \tag{3.2}$$

or a "sellers' market," with constrained or rationed buyers,

$$X = X^S < X^D, \tag{3.3}$$

or a "buyers' market," with constrained or rationed sellers,

$$(3.4) \qquad X = X^D < X^S.$$

Note that without further specification of how the individuals on the long side are rationed, the three properties do not lead to a unique specification of equilibrium. For the remainder of the paper we assume that the pattern of rationing does not affect the level of the effective demands or supplies in the economy.

3.1.3. Effective demand and supply schedules.

We are not considering a single market in isolation: the model includes a flow market for resources, a labour market, and a market for output. These markets are linked by the profit-maximizing representative firm which buys labour services and flows of resource as factor inputs, and sells its produced output to the households, which, as well as earning the wage bill, are also assumed to receive the net profits from industry and the return to owners of resource supply on the resource market. The possibility of trading at "false," or non-market-clearing, prices leads to the possibility of an agent finding himself on the constrained or rationed long side in any market. This in turn may affect his demand or supply schedules on the other markets.

For example, if the representative firm is unable to hire as much labour services as it would like, given the going real wage and real resource price, the consequent rationing may lead it to increase its demand for resource flow as it substitutes one factor input for another. At the same time its supply of output may be reduced. The revised schedules are no longer notional, since notional schedules are formulated assuming balanced, clearing markets with no quantity constraints on any market; rather, the revised schedules are "effective" (Clower (1965)) since they are formulated by taking into account the possibility of quantity constraints if the agent is on the long side of any markets.

There is no reason to expect that the effective demands and supplies of any individual in the three markets will be mutually consistent: in an economy with rationing, *ex ante* supplies and demands are tentative. An individual operating as a rationed buyer on a market will express a demand in this market even if he has little hope that his demand will be fully met. If it is met, he will be able to reconsider his demands and supplies on the other markets. That is, an individual forms his demand or supply schedules separately on each market. In formulating his demand or supply schedule on any market, he does not take into account the rationing he may have to face on this market. But he will not neglect the constraints that limit his trades on other markets. Formalizations of non-market-clearing trading (such as that of Benassy (1973)) attempt to specify fully the constraints on his operations in other markets that are perceived by any individual when he formulates his demand or supply schedules of any good.

As Clower (1965) pointed out, if an economic agent is constrained it is no longer optimal for him to determine all his supply and demand functions at a stroke. Rather he should determine his demand and supply functions one at a time, taking into account in doing so information about other markets. Following Benassy (1973), let the effective demand (supply) of an economic actor on a market be the demand (supply) he will choose by maximizing his utility or profit subject to his budget constraint and to the quantity constraints he perceives on the other markets: he does not take into account the constraints he might experience on the market considered. The rationale for this definition of effective demand is that it is this which is actually transmitted to the market. However, Benassy lists several circumstances when the actual demands transmitted to the market are not equal to the effective demands:

i) if there are transactions costs (monetary or non-monetary) and the economic actor strongly expects to be rationed, he may prefer to stay at home and thus not express any demand at all;

ii) on the other hand, an economic actor expecting severe rationing might place orders far in excess of his actual demand (the effective demand) hoping to receive an amount closer to his actual demand, and knowing that any surplus would be readily resaleable to other rationed buyers.

These cases of distorted demands on the market occur when the economic actor is on the "long" side, heavily rationed, and in particular circumstances. But Benassy argues that the effective demands of the "long" side are not correctly transmitted to the other side of the market. It is thus claimed that the above definition of effective demand (supply) is a good approximation of the actual demands experienced on a market.

The specification used in this model is that the firm formulates its supply of output taking as given the purchases of factor inputs, with the possibility of four supply schedules, depending on whether it is rationed or not on each of the two factor input markets. Similarly, the firm formulates its demand for each factor input taking as given the sales of output and the purchases of the other factor input, with four possible demand schedules. We shall examine the derivation of these schedules more closely below. Note that the firm does not take into account the possibility of being rationed on a market when deriving its demand or supply schedule for that market.

3.2. Types of Fix-Price Equilibria

The nature of the fix-price equilibrium will play an important role in the analysis of impacts of policy measures or of exogenous disturbances. It is thus important to classify the various types of equilibria, to study their respective properties, and to ask which are more likely to occur.

In section 3.1.2 we noted that each of the three markets may be in one of three states: balanced or clearing, a sellers' market, or a buyers' market. Noting that the first state is a boundary between the other two, we find that ignoring the balanced state there are eight (2^3) possibilities for the state of the economy. These possibilities are summarized in Table 3.1, which describes the state of the three markets, the symbols used in this paper, and the labels given to the combination of states of the two markets for output and labour by Malinvaud (1977), who did not consider the possibility of rationing (of either side) in the market for resource flow as a factor input. As shown below, the existence of non-market-clearing trading in the third market can change some policy recommendations, since the behaviour of the system is affected by non-market-clearing in the resource flow market.

The symbols introduced in Table 3.1 are derived from the situation in the three markets of the profit-maximizing representative firm: in

Table 3.1 The eight possibilities.

Output market	Labour market	Resource flow market	Our symbol	Malinvaud's name
sellers' $Y = Y^S < Y^D$ (consumers rationed)	buyers' $N = N^D < N^S$ (workers rationed)	buyers' $R = R^D < R^S$ (suppliers rationed)	SC	classical unemployment
"	"	sellers' $R = R^S < R^D$ (firms rationed)	RC	
"	sellers' $N = N^S < N^D$ (firms rationed)	buyers'	NC	
"	"	sellers'	NRC	repressed inflation
buyers' $Y = Y^D < Y^S$ (firms rationed)	buyers'	buyers'	DC	Keynesian unemployment
"	"	sellers'	DRC	
"	sellers'	buyers'	DNC	
"	"	sellers'	DNRC	uninteresting

each of the three markets the firm can be either on the short, unconstrained or unrationed, side, or on the long, constrained or rationed, side: the short side is that of the sellers if there is excess demand and that of the buyers if excess supply; the long side is that of the buyers if there is excess demand and that of the sellers if excess supply. The assumption of voluntary exchange means that the amount actually traded is the lesser of the amounts supplied and demanded: the amount offered or demanded on the short side of the market is the amount traded.

When the firm is unconstrained on all three markets (it can buy as much of both factor inputs as it wants to, and sell as much output as it wants to, given the going real prices, w and v), the level of activity of the economy is determined by the diminishing returns to scale of the production technology. This is the case SC, for "(output)supply-constrained," of Table 3.1. Neither labour nor the flow of resource is fully employed, and yet firms sell all their output: Malinvaud labels this case "classical unemployment." The output market is a sellers' market and the factor input markets are buyers' markets.

When the firm is unconstrained on the output market but constrained on both factor input markets (it cannot buy as much labour service or resource flow as it would like to, given the going real prices, w and v, but it can sell as much output as it wants to, given w, v, and the supplies of factor inputs, $\bar{N}^S$ and $\bar{R}^S$), the level of activity of the economy is determined by the supplies of factor inputs, $\bar{N}^S$ and $\bar{R}^S$. This is the case NRC, for "labour- and resource-constrained," of Table 3.1. In this case demand exceeds supply on all markets and there is demand-pull inflationary pressure. With the short-run price rigidities there is what Malinvaud calls "repressed inflation." All three markets are sellers' markets.

When the firm is unconstrained on the factor input markets, but contrained on the market for output (it cannot sell as much output as it would like to, given the going real wages w and v, but can buy as much labour services and resource flow as it wants to, given w, v, and the level of aggregate demand for output, Y^D), the level of activity of the economy is determined by the level of aggregate demand for output, Y^D (we ignore the possibility of stockpiling output). This is the case DC, for "(output)demand-constrained," of Table 3.1. In this case there is excess supply on all markets: there is unemployment of labour and resource flow, and firms do not produce more because of lack of aggregate demand for output. Malinvaud labels this case "Keynesian unemployment." All three markets are buyers' markets.

Symmetry suggests a situation where the firm is constrained on all three markets (it cannot sell as much as it would like to, given the going real prices, w and v, and the supplies of factor inputs, $\bar{N}^S$ and $\bar{R}^S$; it cannot buy as much labour service as it would like to, given w, v, $\bar{R}^S$, and the level of aggregate demand for output, Y^D; and it cannot buy as much resource flow as it would like to, given w, v, $\bar{N}^S$, and Y^D). But in this case the level of activity of the economy is determined by at least two of the exogenous variables $\bar{N}^S$, $\bar{R}^S$, and Y^D, since an increase in aggregate demand for output could only be met by an increase in production if the supply of one or both of the factor inputs increased as well. Malinvaud concludes that this case, with a buyers' market for output and sellers' markets for the factor inputs, is not very likely to occur. This is the case DNRC for "(output) demand-, labour-, and resource-constrained," of Table 3.1, throughout which all markets clear, as shown in Appendix B1.

The explicit modelling of a third market, that for resource flow, introduces four additional cases. RC occurs when the firm is unconstrained on the markets for output (a sellers' market) or labour (a buyers' market), but is constrained on the market for resource flow (a sellers' market) where it cannot buy as much resource flow as it would like to, given the real prices, w and v. NC occurs when the firm is unconstrained on the markets for output (a sellers' market) or resource flow (a buyers' market), but is constrained on the market for labour service (a sellers' market) where it cannot buy as much labour services as it would like to, given the real prices, w and v. DRC occurs when the firm is unconstrained on the market for labour (a buyers' market), but is constrained on the markets for output (a buyers' market) and resource flow (a sellers' market): it cannot sell as much output as it would like to, given w, v, and $\bar{R}^S$, and it cannot buy as much resource flow as it would like to, given w, v, and Y^D, but given w, v, Y^D, and $\bar{R}^S$ it can buy as much labour as it wants to. DNC occurs when the firm is unconstrained on the market for resource flow (a buyers' market), but is constrained on the markets for output (a buyers' market) and labour (a sellers' market): it cannot sell as much as it would like to, given w, v, and $\bar{N}^S$, and it cannot buy as much labour service as it would like to, given w, v, and Y^D, but given w, v, Y^D, and $\bar{N}^S$ it can buy as much resource flow as it wants to. Case DNC cannot occur in models with only one variable input factor, since in that case there is a rigid relationship between employment and sales through the production function, which ensures that the firm cannot be constrained on both markets. However, in this model the existence of a second variable factor input breaks this relationship, and this case can occur.

Case RC is similar to SC, the case of classical unemployment: if we ignore the resource market then it is identical, with unsatisfied effective demand on the market for output and unemployment on the labour market. But the firm is constrained--on the market for resource flow, and as we see below in Tables 3.4 and 3.5, the fact that the resource market is a sellers' market alters the behaviour of the economy.

Case NC is similar to NRC, the case of repressed inflation, with unsatisfied effective demand on the markets for output and labour leading to upwards pressure on the money price of output and the money wage. But there is excess effective supply on the market for resource flow, leading to downwards pressure on the money price of resource flow. If there is a fall in the real price of resource flow the level of activity of the economy increases as production increases.

Case DRC is similar to DC, the case of Keynsian unemployment, with excess effective supply on the output market and unemployment on the labour market. But there is unsatisfied effective demand on the resource flow market, and thus no possibility of substitution between labour and resource in response to changes in w or v.

Case DNC is apparently similar to DNRC, the case of effective general-market-clearing, but no market is balanced: there is excess effective supply on the markets for output and resource flow, and full-employment with unsatisfied demand on the labour market. All quantities traded are inelastic.

3.3. SC: The case of classical unemployment.

This section analyzes the determination of output, employment, and resource flow when the values of P, W, and V are such that excess demand exists in the market for output, and excess supply exists in the markets for labour services and resource flow. The economy is "(output) supply-constrained," hence SC. In this situation the principle of voluntary exchange implies that output is supply-determined and that employment and resource use are both demand-determined. The firm is unconstrained on all three markets, and the level of activity of the economy is determined by the diminishing returns to scale of the production technology at any combination of real prices, w and v. We assert that

$$(3.5)\qquad \begin{aligned} N &= N^D(w, v) < \bar{N}^S \\ R &= R^D(w, v) < \bar{R}^S \\ Y &= Y^S(w, v) < Y^D \end{aligned}$$

and we argue that quantities determined in this way are not inconsistent. Written as functions of w and v only, these functions are the same notional functions that we introduced in section 2.2.1 in equation (2.11), although the economy is no longer balanced, and no market clears.

3.3.1. The behaviour of the firms in the SC case.

In the basic model of Chapter II, the representative firm maximized (equation (2.3))

$$(3.6)\qquad \pi = Y^S - wN^D - vR^D$$

subject only to a given real price combination (w, v) and to the constraint of the production function (equations (2.5) and (2.6))

$$(3.7) \qquad Y^S = F(N^D, R^D).$$

Underlying this formulation is the assumption that the representative firm could sell all the output which it offered for sale and could buy all the labour and resource which it demanded at the going real price vector (w, v). Since the representative firm is on the short side in all three markets, this assumption holds for the SC case, and the derived functions of equation (2.11) are those of the SC case:

$$(3.8) \qquad N^D = N^{SCD}(w, v)$$

$$R^D = R^{SCD}(w, v)$$

$$Y^S = Y^{SCS}(w, v).$$

These functions are derived from solution of the first-order conditions for profit-maximization, as in Appendix B1.

Standard analysis of Appendix B1 yields the partial derivatives of equation 2.12 and Table 3.3. We see that for case SC, the case of classical unemployment, that the output supply function and both factor input demand functions are decreasing functions of both the real wage and the real resource flow price.

3.3.2. The behaviour of the households in the SC case.

From discussion in section 2.3.1 the *ex ante* consumption function is given by equation (2.24):

$$(3.9) \qquad Y^D = (1 - s)(Y^S - T + w(\bar{N}^S - N^D) + v(\bar{R}^S - R^D)) + I + G$$

since the *ex ante* disposable income is given by

(3.10) $$Y^{DI} = w\bar{N}^S + v\bar{R}^S + \pi - T$$

where the *ex ante* profits are given by

(3.11) $$\pi = Y^S - wN^D - vR^D$$

and since the aggregate demand for output is given by

(3.12) $$Y^D = (1-s)Y^{DI} + I + G.$$

But *ex post*, the actual levels of transactions on the three markets are given by the short sides of the markets. In the SC case these are $Y = Y^{SCS}$, $N = N^{SCD}$, and $R = R^{SCD}$. Thus the *ex post*, actual, amounts for the wage-bill, the resource-bill, and the profits are wN^{SCD}, vR^{SCD}, and $Y^{SCS} - wN^{SCD} - vR^{SCD}$, respectively. Thus the *ex post*, actual, disposable income is

(3.13) $$Y^{DI} = Y^{SCS} - T$$

and the *ex post*, actual, aggregate demand for output is

(3.14) $$Y^D = (1-s)(Y^{SCS} - T) + I + G > Y^{SCS}.$$

Then,

(3.15) $$Y^D_w = (1-s)Y^{SCS}_w < 0$$

$$Y^D_v = (1-s)Y^{SCS}_v < 0.$$

We note that as the real prices decrease Y^{SCS} increases, and with it but less rapidly Y^D until the two are equal, or the output market clears. This behaviour will be analyzed more closely in section 3.4.1 below.

3.3.3. The determination of quantities in the SC case.

As analysed above, the SC case of classical unemployment, where the output market is a sellers' market and the markets for factor inputs are both buyers' markets, is the case where the firm's notional schedules of section 2.2.1 are also applicable with non-market-clearing trading, since the firm is on the short side in all three markets.

The principle of voluntary exchange means that the level of output in the case of classical unemployment is the output ($Y = Y^{SCS}$) corresponding to the employment ($N = N^{SCD}$) at which the marginal product of labour (F_N) equals the real wage (w), and to the resource use ($R = R^{SCD}$) at which the marginal product of resource flow (F_R) equals the real price of resource flow (v).

In the market for output, the excess demand is positive:

$$(3.16) \qquad Y^X = Y^D - Y^{SCS} = (1-s)(Y^{SCS} - T) + I + G - Y^{SCS} > 0$$

where, from equation (3.8),

$$Y^{SCS} = F(N^{SCD}, R^{SCD}) = Y^{SCS}(w, v).$$

In the market for labour, the excess demand is negative:

$$(3.17) \qquad N^X = N^{SCD}(w, v) - \bar{N}^S < 0,$$

as is the excess demand in the market for resource flow:

$$(3.18) \qquad R^X = R^{SCD}(w, v) - \bar{R}^S < 0.$$

That is, there is unfulfilled demand for output, unemployment, and unused flow of resource.

3.3.4. Comparative statics of the SC case.

Increased government expenditure G would not reduce the unemployment level or increase the use of resource, nor would it increase the flow output since the level of output is technology-determined by the decreasing returns to scale of the production process. However, a reduction in the real wage w or the real resource price v or both would lead to a reduction in unemployment and an increase in resource use as the level of output increased.

Changes in the supply of labour $\bar{N}^S$ or the supply of resource flow $\bar{R}^S$ would have no effect on the fix-price non-market-clearing equilibrium of the SC case, since both factor input markets are buyers' markets. But an improvement in production technology would affect the level of activity of the economy. Appendix A3 shows that in the case of neutral technical change

$$(3.19) \qquad Y = F^2(N, R) \equiv \alpha F(N, R), \qquad \alpha > 1$$

the demand for both factor inputs would increase. Since both factor input markets are buyers' markets in the SC case, voluntary exchange means that the amounts of labour services and resource flow traded would increase and so the level of unemployment fall. The level of output would increase.

Appendix A3 shows that in the case of resource-augmenting technical change

$$(3.20) \qquad Y = F^1(N, R) \equiv F(N, \alpha R), \qquad \alpha > 1$$

the demand for labour would increase, and so the level of unemployment would fall, since the labour market is a buyers' market. But the change in the demand for resource flow is unclear:

Appendix A3 shows that for large α it would be negative and in this case the level of resource use would fall. The level of output would increase.

3.4. DC: The case of Keynesian unemployment.

This section analyses the determination of output, employment, and resource flow when the values of P, W, and V are such that excess supply exists on all three markets, which are all buyers' markets. The firm is "(output) demand constrained," hence DC. In this situation the principle of voluntary exchange implies that employment, resource use, and output are all demand-determined, given that firms adjust output to equal sales at all times with no inventories of output. But this does not mean that the notional functions will determine the quantities,

$$N = N^{SCD} < \bar{N}^S,$$

$$R = R^{SCD} < \bar{R}^S,$$

$$Y = Y^D < Y^{SCS} = F(N^{SCD}, R^{SCD}),$$

since quantities determined in this manner would be inconsistent. In particular, if firms were actually constrained to produce output less than Y^{SCS}, their demands for factor inputs would not both be given by the notional functions $N^{SCD}(w, v)$ and $R^{SCD}(w, v)$; while if there is excess supply on the market for output, we show in section 3.4.1 below that aggregate demand for output is not given by equation (3.14) of section 3.3.2 above. The rest of this chapter is concerned with resolving the problem that the notional schedules are derived with the assumption that each agent can sell all offers or buy all demands at the going prices: notional schedules fail to take account of the fact that one market's failure to clear creates a constraint which will influence behaviour on other markets.

3.4.1. The behaviour of the households in the DC case.

From the discussion in sections 2.3.1 and 3.3.2 the *ex post*, actual, disposable income is

(3.22) $$Y^{DI} = Y - T,$$

where voluntary exchange implies that the level of output is given by

(3.23) $$Y = \min(Y^D, Y^S).$$

Thus the *ex post*, actual, aggregate demand for output is

(3.24) $$Y^D = (1-s)(Y-T) + I + G$$

with two cases to consider:

(i) If $Y^S < Y^D$, then the market for output is a sellers' market with the buyers of output (the households) rationed. This includes the regions SC, RC, NC, and NRC in Table 3.1, and, as in section 3.3.2, aggregate demand for output is given by

(3.25) $$Y^D = (1-s)(Y^S - T) + I + G > Y^S.$$

(ii) If $Y^D < Y^S$, then the market for output is a buyers' market with sellers of output (the firms) rationed. This includes the regions DC, DRC, DNC, and DNRC in Table 3.1. In this case, including DC, the case of Keynesian unemployment, aggregate demand for output is given by

$$(3.26) \qquad Y^D = (1-s)(Y^D - T) + I + G < Y^S,$$

which can be solved for Y^D to give

$$(3.27) \qquad Y^D = \bar{Y} \equiv (I + G - (1-s)T)/s.$$

This states that when the output market is a buyers' market ($Y^D < Y^S$), aggregate demand for output Y^D is a constant, $\bar{Y}$.

It is possible to plot Y^D against Y^S to obtain the disequilibrium version of the Keynesian cross diagram. This is done in Figure 3.1. From above

$$(3.28) \qquad Y_i^S < Y_i^D < \bar{Y} = Y_{ii}^D < Y_{ii}^S$$

where the subscripts i and ii refer to the cases above (and the regions in Figure 3.1). The two regions in Figure 3.1 should be compared with the traditional Keynesian cross diagram of Figure 2.1, in which

$$Y^D = (1-s)(Y^S - T) + I + G$$

In Figure 3.1,

$$(3.29) \qquad Y^D = \begin{cases} (1-s)(Y^S - T) + I + G, & Y^D > Y^S, \text{ (region (i))} \\ \bar{Y} \equiv (I + G - (1-s)T)/s, & Y^D < Y^S, \text{ (region (ii))}. \end{cases}$$

Thus, when $Y^D > Y^S$ (in region (i)), the traditional and the disequilibrium approaches are identical; when $Y^D < Y^S$ (in region (ii)), the traditional and the disequilibrium approaches differ, with inelastic aggregate demand.

We are concerned with the elasticity (which has the sign of the partial derivative) of excess demand in the two regions, where excess demand Y^X is given by

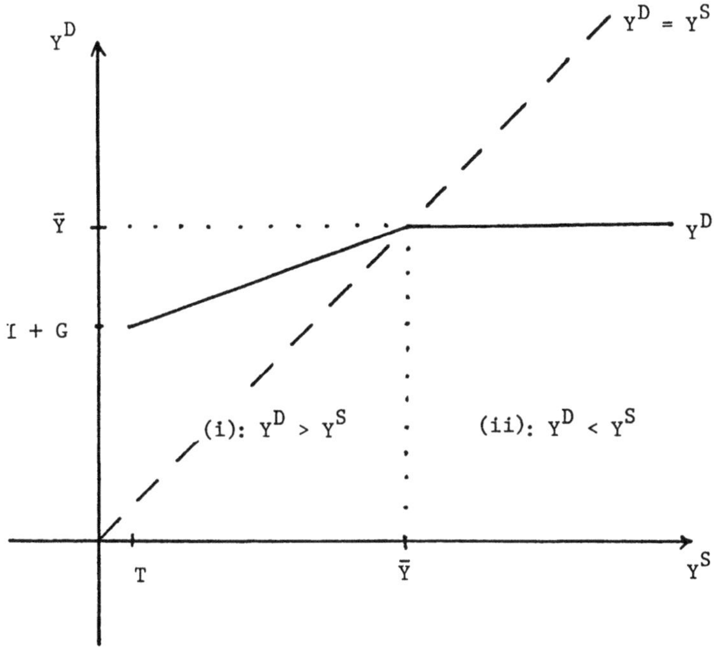

Figure 3.1 The disequilibrium Keynesian cross

$$(3.30) \qquad Y^X \equiv Y^D - Y^S.$$

In region (i), the areas SC, RC, NC, and NRC,

$$(3.31) \qquad Y^X = (Y^S - T)(1-s) + I + G - Y^S = s(\bar{Y} - Y^S) > 0$$

$$(3.32) \qquad Y^X_w = -s\, Y^S_w$$

$$Y^X_v = -s\, Y^S_v \,.$$

In region (ii), the areas DC, DRC, DNC, and DNRC,

$$(3.33) \qquad Y^X = \bar{Y} - Y^S < 0$$

$$(3.34) \qquad Y^X_w = -Y^S_w$$

$$Y^X_v = -Y^S_v \,.$$

Thus we see that the excess demand for output is a non-decreasing function of both the real wage and the real resource flow price, for inelastic factor supplies. Since $0 < s < 1$, the excess demand for output is less elastic in region (i) than in region (ii).

3.4.2. The behaviour of the firms in the DC case.

From the preceding section we see that in this case, actual level of sales of output (and hence actual production with no stockpiling) is given by $\bar{Y}$, which is less than the notional supply Y^{SCS}. Given $Y < Y^{SCS}$, the firm acts as a quantity taker with respect to its sales, as well as acting as a real wage and real resource price taker. In the basic model of section 2.2.1 and in the SC case of section 3.3.1 the level of sales was a choice variable. Now, however, the level of sales is a demand-determined constraint on its sales. Formally, the representative firm has to choose its effective demands for labour services and resource flow, represented by N^{DCD} and R^{DCD}, respectively, so as to maximize its profit

$$(3.35) \qquad \pi = Y - wN^{DCD} - vR^{DCD} \qquad \text{subject to}$$

$$Y = F(N^{DCD}, R^{DCD}),$$

$$Y \leq \bar{Y}.$$

This problem is solved in Appendix B1, to yield the effective demand schedules in terms of w, v, and $\bar{Y}$:

$$(3.36) \qquad N^D = N^{DCD}(w/v, \bar{Y})$$

$$R^D = R^{DCD}(w/v, \bar{Y})$$

such that

$$(3.37) \qquad \bar{Y} = F(N^{DCD}, R^{DCD}),$$

and that the quantities determined are consistent.

The partial derivatives of the effective demand schedules are derived in Appendix B2 and listed in Table 3.3. It is seen that the own real price elasticities are negative, and the cross real price elasticities positive as substitution of input factors occurs. Not surprisingly, demand for both factor inputs increases as the level of autonomous demand $\bar{Y}$ increases.

In section 3.3.1 notional demands for factor inputs were derived as functions of the two real prices, but not of output, which was maximized out as a separate choice variable. In contrast, the effective demand schedules derived above are functions of output. The level of demand for goods here imposes output, which equals sales, as a constraint on the effective demands for labour and resource flow. One implication of the effective schedules N^{DCD} and R^{DCD} is that the effective demand for factor inputs can vary even with the real prices fixed: changes in the level of constraint $\bar{Y}$ influence the effective demands independently of changes in w or v.

As shown in Appendix B1, the demand constraint on sales causes the representative firm to operate in a region (shown as DC in Figure 3.3a) where the marginal product of labour exceeds the real wage and the marginal product of resource flow exceeds the real resource flow price. (Note that the assumption of voluntary exchange means that the firm will never operate in a region where the real prices exceed their respective marginal products, since that would require $Y > Y^{SCS}(w, v)$.) If the constraint on sales were eased, the representative firm would respond, at the existing real price vector (w, v), by raising output, employment, and resource use. If possible, this process would continue until the marginal products fell sufficiently to equal their respective real prices, at which point the representative firm would be operating according to its notional supply of output and demands for input factors. Note that along the profit-maximizing expansion path the ratio of the input factor marginal products equals the ratio of their respective real prices:

(3.38) $$F_N/F_R = w/v.$$

It is possible to plot the labour market in this region, in a figure similar to Figure 2.2 of the basic model. Such a plot is shown in Figure 3.2, which reproduces an N^D curve from Figure 2.2, here labelled as N^{SCD}, the notional demand for labour, against the real wage. The curve N^{DCD} depicts the effective demand for labour derived above. For a given value of $\bar{Y}$, N^{DCD} coincides with N^{SCD} when the combination of (w, v) is such that $\bar{Y}$ is not an operative constant, that is, when $Y^{SCS} \leq \bar{Y}$. When (w, v) is such that $\bar{Y}$ is a binding constraint, that is, when $\bar{Y} < Y^{SCS}$, then N^{DCD} is independent of the two real prices as long as their ratio w/v is constant. It is here that such a diagram as Figure 3.2 is misleading, since the demand for labour against the real wage has been plotted ceteris paribus, and the non-zero elasticity of N^{DCD} represents only substitution from resource to labour as the real wage falls, for constant real resource price. Plotting labour demand against the real resource price would reveal an upwards sloping N^{DCD} segment as substitution occurred from resource to labour as the real resource price fell, for constant real wage. Figure 3.3, in the (w, v)-plane, will give a better idea of the model.

3.4.3. The determination of quantities in the DC case.

As analysed above, in the DC case of Keynesian unemployment, with excess supply on all three markets, the quantities of employment, resource use, and output traded are all demand-determined. The level of output is given by the exogenous demand ($Y = \bar{Y}$) and is produced by the employment ($N = N^{DCD}$) and resource use ($R = R^{DCD}$) that satisfy the production function ($\bar{Y} = F(N^{DCD}, R^{DCD})$), and at which levels the ratio

(a)

(b)

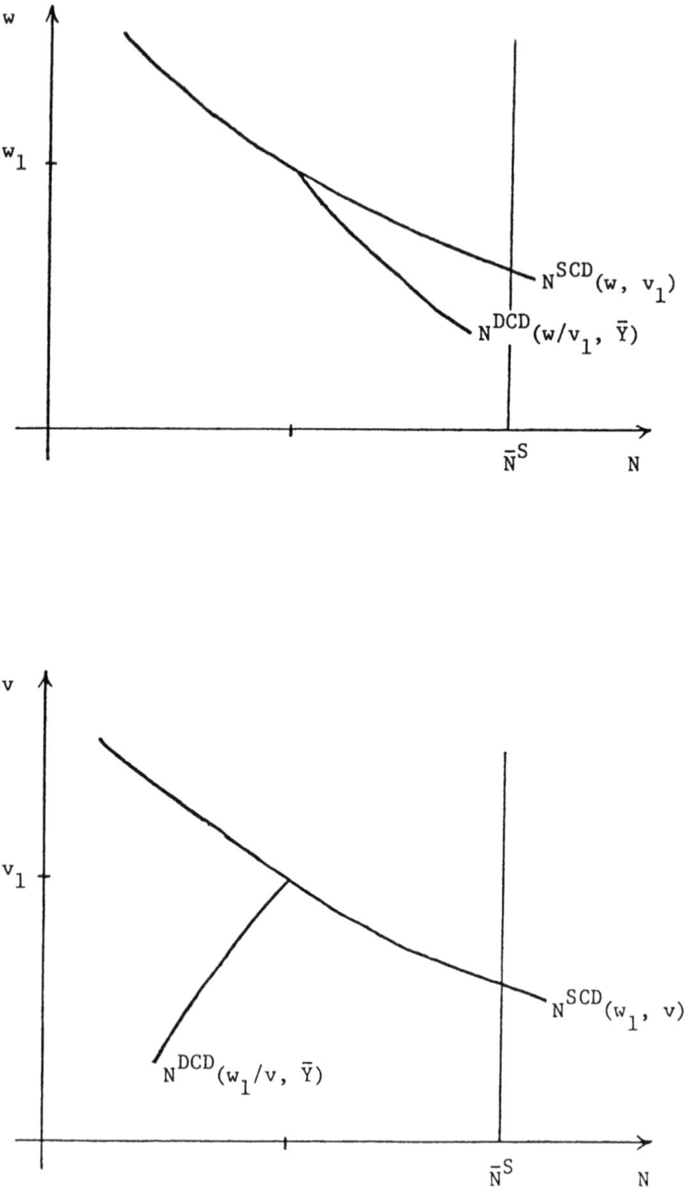

Figure 3.2: The labour market in the DC case.

of the marginal products equals the ratio of the real prices (F_R/F_N = w/v).

Note that this case can only occur when the level of autonomous demand is less than full-employment output:

$$(3.39) \qquad \bar{Y} < F(\bar{N}^S, \bar{R}^S).$$

It is seen that DC is the non-market-clearing trading equivalent of the generalized excess supply region V of Figure 2.10a and Table 2.1.

In the case of Keynesian unemployment there is excess supply on all three markets. In the market for labour

$$(3.40) \qquad N^X = N^{DCD} - \bar{N}^S < 0.$$

In the market for resource flow

$$(3.41) \qquad R^X = R^{DCD} - \bar{R}^S < 0.$$

In the market for output, we want to formulate the supply of output, which is on the long side of the market since the firm is constrained in its sales of output. We use the Benassy assumption that an agent will not take into account the possibility of being constrained on the market considered in formulating his demand or supply schedule for that market. Hence the effective supply of output will be $Y^{SCS}(w, v)$, which is derived assuming that the firm can sell as much output as it wants to, at the going real prices, w and v. The excess demand in the market for output is given by

$$(3.42) \qquad Y^X = \bar{Y} - Y^{SCS} < 0.$$

That is, there is unemployment of labour and resource flow, and firms would like to produce and sell more output.

3.4.4. Comparative statics of the DC case.

Increased government expenditure G would lead to an increase in aggregate demand for output, since

$$(3.43) \qquad Y^D = \bar{Y} \equiv (I + G - (1-s)T)/s,$$

and since the market for output is a buyers' market, the level of production of the economy would increase to meet the increased demand, and there would be an increase in employment and resource use. A reduction in the household propensity to save s would have a similar effect, as the household demand for output increased.

Changes in the supply of labour $\bar{N}^S$ or the supply of resource flow $\bar{R}^S$ would have no effect on the fix-price non-market-clearing equilibrium of the DC case, since both factor input markets are buyers' markets. Changes in the ratio of real prices, w/v, although they cannot affect the level of output of the economy which is demand-determined, would affect the levels of employment and resource use by inducing substitution between labour and resource flow in the production process. It is noteworthy that a fall in v can lead to an increase in unemployment as resource becomes more attractive an input to the representative firm.

Neutral technical change would have no effect on the level of output, which is demand-determined, but would lead to a reduction in the level of employment and the level of resource use, as shown in Appendix B3.

Resource-augmenting technical change

$$(3.44)\qquad Y = F^2(N, R) \equiv F(N, \alpha R), \qquad \alpha > 1,$$

would not affect the level of output of the economy, but would reduce the level of employment and, for large α, reduce the level of resource use, as shown in Appendix B3.

3.5. NRC: the case of repressed inflation.

This section analyzes the determination of output, employment, and resource use when the values of P, W, V, and $\bar{Y}$ are such that excess demand exists on all three markets, which are sellers' markets. The firm is on the long side in the two factor input markets, that is, it is "labour- and resource-constrained", hence NRC. In this situation the principle of voluntary exchange implies that the quantities are all supply-determined. For the factor inputs, the quantities are

$$(3.45)\qquad N = \bar{N}^S < N^{SCD}$$

$$R = \bar{R}^S < R^{SCD}$$

but the principle does not mean that

$$Y = Y^{SCS} < Y^D = \bar{Y}$$

since, just as the notional demand functions were not relevant in the case of generalized excess supply, so the notional supply function for output, Y^{SCS}, is not relevant in the case of generalized excess demand. The output determined by the notional supply of output function would be inconsistent with the quantities of inputs used. More particularly, if the firms were constrained to buy less than their notional demands on both factor input markets, their supply of output would not be given by the notional supply function Y^{SCS}. Output would be reduced.

3.5.1. The behaviour of the households in the NRC case.

The output market is a sellers' market with the representative firm able to sell as much as it would like to, given w, v, $\bar{N}^S$, and $\bar{R}^S$. That is, on the market for output there is excess demand. From the discussion in section 3.4.1, NRC, the case of repressed inflation, falls into case (i), with

$$(3.46) \qquad Y^D = (1-s)(Y^S - T) + I + G > Y^S$$

and

$$(3.47) \qquad Y^X = (1-s)(Y^S - T) + I + G - Y^S = s(\bar{Y} - Y^S) > 0$$

3.5.2. The behaviour of the firms in the NRC case.

Excess demand in the labour market means that the representative firm will not be able to satisfy its demand for labour, formulated given w, v, and $\bar{R}^S$. Excess demand in the resource flow market means that the representative firm will not be able to satisfy its demand for resource, formulated given w, v, and $\bar{N}^S$. Voluntary exchange implies that actual purchases will equal the quantity supplied in each of the markets, that is,

$$(3.48) \qquad N = \bar{N}^S \quad \text{and} \quad R = \bar{R}^S.$$

Given $N < N^{RCD} < N^{SCD}$ and $R < R^{NCD} < R^{SCD}$, where N^{RCD} is the labour demand function given full use of resource flow and where R^{NCD} is the resource demand function given full employment, the representative firm acts as a quantity taker with respect to employment and resource use, as well as acting as a real prices taker. The levels of factor inputs purchases are no longer choice variables; they are now supply-constrained. Profit-maximization now implies producing as much as possible with the available labour and resource flow. The maximum quantity is written as Y^{NRCS}, where

$$(3.49) \qquad Y^{NRCS} = F(\bar{N}^S, \bar{R}^S).$$

Obviously this case can only occur when the level of autonomous demand for output is greater than full-employment output, that is, when

$$(3.50) \qquad \bar{Y} > Y^{NRCS} = F(\bar{N}^S, \bar{R}^S).$$

Appendix B1 shows that, as in the Keynesian unemployment case DC in section 3.4, the representative firm is forced to operate in a region where the marginal product of each of the input factor exceeds its respective real price.

The notional supply of output Y^{SCS}, derived in Chapter II and section 3.3.1, is a function of the two real prices, w and v, but not of the supply of factor inputs. Employment and resource use were maximized out as separate choice variables. In contrast, the effective supply of output Y^{NRCS} is a function of the levels of employment and resource use, with the levels of labour supply and resource supply constraining the effective supply of output. Y^{NRCS} implies that the effective supply of output can vary even with constant real prices. Changes in the levels of the constraints $\bar{N}^S$ and $\bar{R}^S$ affect output supply independently of changes in w or v.

3.5.3. The determination of quantities in the NRC case.

As analysed above, in the NRC case of repressed inflation, with excess demand on all three markets, the quantities of employment, resource use, and output traded are all supply-determined. The level of output is the full-employment output ($Y = F(\bar{N}^S, \bar{R}^S)$) and is produced by full-employment of both input factors ($N = \bar{N}^S$ and $R = \bar{R}^S$).

Note that this case can only occur when the level of autonomous demand is greater than full-employment output:

$$\bar{Y} > F(\bar{N}^S, \bar{R}^S),$$

in contrast to the DC case of Keynesian unemployment in section 3.4 above which could only occur when the level of autonomous demand was <u>less</u> than full-employment output. Note that NRC is the non-market-clearing trading equivalent of the generalized excess demand region IV of Figure 2.10b and Table 2.1.

In the case of repressed inflation there is excess demand on all three markets. In the market for output

$$(3.51) \qquad Y^X = Y^D - Y^{NRCS} = s(\bar{Y} - Y^{NRCS}) > 0$$

In the two factor-input markets we formulate the demands, which are on the long side of the market, by using the Benassy assumption that in formulating his demand or supply schedule for a market, an agent will consider all constraints but the possible constraint on that market. Hence the effective demand for labour is formulated assuming a sellers' market for resource flow, $N^{RCD}(w, \bar{R}^S)$, and the effective demand for resource is formulated assuming a sellers' market for labour, $R^{NCD}(v, \bar{N}^S)$.

(These two functions are derived in Appendix B1.) The excess effective demands are

$$(3.52) \qquad N^X = N^{RCD} - \bar{N}^S > 0,$$

$$R^X = R^{NCD} - \bar{R}^S > 0.$$

That is, firms would like to buy more of both factor inputs.

3.5.4. Comparative statics of the NRC case.

Changes in the level of autonomous demand for output $\bar{Y}$ would have no effect on the fix-price non-market-clearing equilibrium of the NRC case, since the market for output is a sellers' market. So changes in government expenditure, or taxation, or the household propensity to save would have no effect on the economy.

Since the supplies of factor inputs are inelastic, changes in the real prices w and v would have no effect on the equilibrium. But an increase in the labour supply $\bar{N}^S$ or the resource flow $\bar{R}^S$ would lead to an increase in employment or resource use, respectively, and so to an increase in output. Output would similarly increase with technical improvement of either kind, because the representative firm would be able to make better use of the existing factor input supplies in consequence.

3.6. Output, employment and resource use in general.

This section is concerned with making more general the analysis of the determination of the quantities traded under fix-price non-market-clearing conditions. Any particular combination of real prices implies for each market either that the quantity supplied exceeds the quantity demanded or that the quantity demanded exceeds the quantity supplied or that these two quantities are equal. As noted in section 3.2, there are thus eight possible cases of non-market-clearing, although, depending on the size of the autonomous demand for output $\bar{Y}$ compared to the full-employment output $F(\bar{N}^S, \bar{R}^S)$, not all of them can be attained. These eight cases were introduced in section 3.2 and Table 3.1. So far we have examined three cases in detail: SC, DC, and NRC.

The effective supply schedules for output and effective demand schedules for labour and resource flow are derived for all eight cases in Appendix B1, and summarized in Table 3.2, which incorporates the Benassy assumption of ignoring the possible constraint on any market when deriving the effective schedule for that market. This means, for example, that in the DRC case, although the firm is on the long side in the resource flow market and cannot buy more than $\bar{R}^S$, its derived effective demand for resource flow is given by $R^{DCD}(w/v, \bar{Y})$, which ignores the constraint on purchases in the resource market, but takes account of the constraint on sales in the output market. The signs of the partial derivatives of the effective schedules with respect to the two real prices w and v are derived in Appendix B1 and summarized in Table 3.3.

Having made the Benassy assumption of perceived constraints in deriving the effective schedules, we can proceed to use the assumption of voluntary exchange to derive the actual quantities traded for the eight regions. These are summarized in Table 3.4. Table 3.5 shows the signs of the elasticities of the quantities traded for the eight regions, derived using Tables 3.3 and 3.4. Note that in region DC it is possible

Table 3.2. Effective schedules in the eight regions.

Region	Y^S	N^D	R^D
SC	$Y^{SCS}(w, v)$	$N^{SCD}(w, v)$	$R^{SCD}(w, v)$
RC	$Y^{RCS}(w, \bar{R}^S)$	$N^{RCD}(w, \bar{R}^S)$	R^{SCD}
NC	$Y^{NCD}(v, \bar{N}^S)$	N^{SCD}	$R^{NCD}(v, \bar{N}^S)$
NRC	$Y^{NRCS}(\bar{N}^S, \bar{R}^S)$	N^{RCD}	R^{NCD}
DC	Y^{SCS}	$N^{DCD}(w/v, \bar{Y})$	$R^{DCD}(w/v, \bar{Y})$
DRC	Y^{RCS}	$N^{DRCD}(\bar{R}^S, \bar{Y})$	R^{DCD}
DNC	Y^{NCS}	N^{DCD}	$R^{DNCD}(\bar{N}^S, \bar{Y})$
DNRC	Y^{NRCS}	N^{DRCD}	R^{DNCD}

Table 3.3. Signs of the elasticities of the effective schedules.

Region	Y^S_w	Y^S_v	N^D_w	N^D_v	R^D_w	R^D_v
SC	-	-	-	-	-	-
RC	-	0	-	0	-	-
NC	0	-	-	-	0	-
NRC	0	0	-	0	0	-
DC	-	-	-	+	+	-
DRC	-	0	0	0	+	-
DNC	0	-	-	+	0	0
DNRC	0	0	0	0	0	0

Table 3.4. Actual quantities traded in the eight regions.

Region	Output, Y	Employment, N	Resource use, R
SC	$Y^{SCS} < Y^D$	$N^{SCD} < \bar{N}^S$	$R^{SCD} < \bar{R}^S$
RC	$Y^{RCS} < Y^D$	$N^{RCD} < \bar{N}^S$	$\bar{R}^S < R^{SCD}$
NC	$Y^{NCS} < Y^D$	$\bar{N}^S < N^{SCD}$	$R^{NCD} < \bar{R}^S$
NRC	$Y^{NRCS} < Y^D$	$\bar{N}^S < N^{RCD}$	$\bar{R}^S < R^{NCD}$
DC	$\bar{Y} < Y^{SCS}$	$N^{DCD} < \bar{N}^S$	$R^{DCD} < \bar{R}^S$
DRC	$\bar{Y} < Y^{RCS}$	$N^{DRCD} < \bar{N}^S$	$\bar{R}^S < R^{DCD}$
DNC	$\bar{Y} < Y^{NCS}$	$\bar{N}^S < N^{DCD}$	$R^{DNCD} < \bar{R}^S$
DNRC	$\bar{Y} = Y^{NRCS}$	$\bar{N}^S = N^{DRCD}$	$\bar{R}^S = R^{DNCD}$

Table 3.5. Signs of the elasticities of the quantities traded.

Region	Y_w	Y_v	N_w	N_v	R_w	R_v
SC	-	-	-	-	-	-
RC	-	0	-	0	0	0
NC	0	-	0	0	0	-
NRC	0	0	0	0	0	0
DC	0	0	-	+	+	-
DRC	0	0	0	0	0	0
DNC	0	0	0	0	0	0
DNRC	0	0	0	0	0	0

to reduce unemployment (to increase N) by increasing the real price of resource flow v. Also note that in over half of the regions, lowering the real wage w will not result in any increase of employment.

The section below considers the problem of associating each of the combinations of real prices (w, v) with one of the eight possible non-market-clearing cases, or with one of the cases in which one (or more) of the markets clears, given the level of autonomous demand $\bar{Y}$ and the full-employment output $F(\bar{N}^S, \bar{R}^S)$.

3.6.1. <u>Effective market-clearing loci.</u>

In Figures 2.5 and 2.6 the loci labelled $N^D = \bar{N}^S$, $R^D = \bar{R}^S$, and $Y^D = Y^S$ depicted combinations of the real prices w and v which were consistent with equality between the notional demands and supplies for labour services, resource flow, and output, respectively. An implication of the analysis in the present chapter is that these loci describe market-clearing real price vectors only under the assumption of recontracting. Section 3.4 showed that, when actual trading is taking place with excess supply in the output market and in either one of the factor input markets, the notional demand function in the other factor input market becomes irrelevant, and is replaced by an effective demand function. Similarly, section 3.5 showed that when actual trading is taking place with excess demand on the two factor input markets, the notional supply function in the output market becomes irrelevant, and is replaced by an effective supply schedule. That is, the three notional demand and supply schedules are relevant only when all three markets clear simultaneously, which occurs only at point NRD in Figure 2.5, although we can use the notional schedules in region SC in which the firm is on the short side of all three markets and hence unconstrained.

It is possible to plot the effective market-clearing loci in the two dimensions of real wage w and real resource price v. As in Chapter II, the loci will vary depending on whether the autonomous demand for output $\bar{Y}$ is less than, or equal to, or greater than, the full-employment output $F(\bar{N}^S, \bar{R}^S)$. Figure 3.3a has been plotted for the case

(3.53) $$\bar{Y} < F(\bar{N}^S, \bar{R}^S).$$

The line R—DR—O in Figure 3.3a is the effective locus of market clearing for the resource flow market: for given aggregate demand, and inelastic supplies of labour and resource flow, the resource flow market clears for any pair of real wage and real resource price along the locus. The locus divides the positive quadrant of the (w, v)-plane into two regions: to the left, for lower values of real resource price v, there is excess demand in the resource flow market, that is, the firm finds itself on the long side as it demands resource flow as a factor input--it is constrained by the supply; to the right, at higher values of v, there is excess supply in the resource flow market, and the firm is on the unconstrained, short side.

The line N—DN—O is the effective locus of market clearing for the labour market. Below it, at lower values of real wage w, there is excess demand in the labour market and the firm is on the constrained, long side while there is full employment of labour; above it, at higher values of w, there is excess supply in the labour market, or unemployment

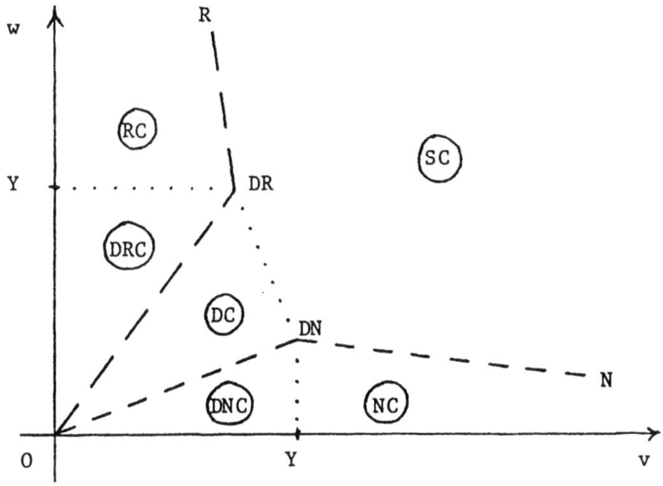

Figure 3.3a: The effective market-clearing loci, $\bar{Y} < F(\bar{N}^S, \bar{R}^S)$.

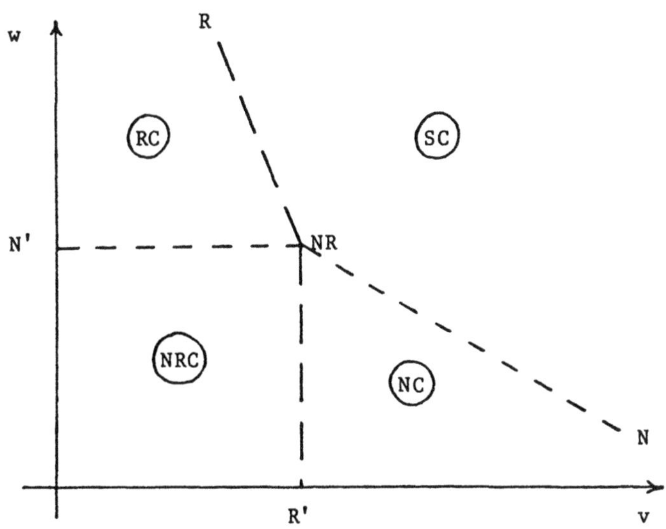

Figure 3.3b: The effective market-clearing loci, $\bar{Y} > F(\bar{N}^S, \bar{R}^S)$.

of labour, and the firm is on the unconstrained, short side while the sellers of labour, the households, are on the constrained, long side.

The line Y—DR—DN—Y is the effective locus of market clearing in the market for output. At values of (w, v) to its northeast, there is excess demand for output: the sellers of output, the firms, are on the unconstrained, short side of the market, and the buyers of output, the households, are on the constrained, long side. At values of (w, v) to the southwest of the locus, there is excess supply of output: the firms are on the constrained, long side while the households are on the unconstrained, short side.

At the point of intersection of the resource and output market-clearing loci, DR, there is clearing on these two markets, while the third, that for labour, is in a state of excess supply. At the point of intersection of the labour and output market-clearing loci, DN, there is clearing on these two markets, while the third, that for resource flow, is in a state of excess supply.

Excess supply of output is only possible if the supplies of factor inputs are sufficiently large and the aggregate demand for output sufficiently small that the demand can be satisfied by the supply of output. Figure 3.3a has been drawn with this assumption:

$$\bar{Y} < F(\bar{N}^S, \bar{R}^S).$$

If the supplies of factor inputs are too small or the autonomous demand too large, then Figure 3.3a will no longer describe the situation. Figure 3.3b has been drawn for the assumption

$$\bar{Y} > F(\bar{N}^S, \bar{R}^S). \tag{3.54}$$

Again there are loci of market-clearing, but since there is excess demand for output even when factor inputs are fully employed, there can be no locus for market-clearing in the output market. The line R—NR—R' is the effective locus of market-clearing in the market for resource flow: to the left there is excess demand, to the right, excess supply. The line N—NR—N' is the effective locus of market-clearing in the market for labour: below it there is full employment with excess demand for labour. At the point of intersection of the labour and resource market-clearing loci, NR, there is clearing on these two markets, while the third, that for output, is in a state of excess demand.

Only when, perhaps because of government fiscal action, the autonomous demand for output is equal to full-employment output,

$$(3.55) \qquad \bar{Y} = F(\bar{N}^S, \bar{R}^S),$$

will a point occur when all three markets clear. Figure 3.4 has been plotted to show this case. The line R—DNR is the effective locus of market-clearing in the market for resource flow: to the left there is excess demand, to the right, excess demand. The line N—DNR is the effective locus of market-clearing in the market for labour: below it there is full employment with excess demand for labour; above it, unemployment with excess supply of labour. Throughout region DNRC there is clearing in all three markets: it is a region of effective general-market-clearing which can only occur if all the exogenous variables are consistent, with the autonomous demand equal to full-employment output. In the rest of the plane, regions SC, RC and NC, there is excess demand on the market for output. There is nowhere excess supply on the market for output.

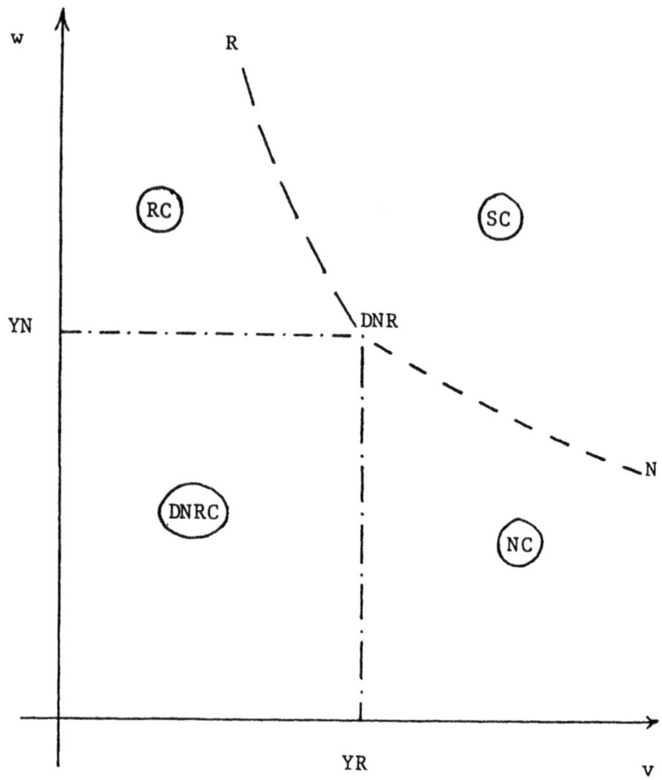

Figure 3.4: The effective market-clearing loci, $\bar{Y} = F(\bar{N}^S, \bar{R}^S)$.

It is possible to give a more intuitive derivation of the patterns of loci. From the discussion of SC, if w and v are initially very high, the firm will operate in the SC case. As the real wage w is reduced, holding v constant the factor demands and the output supply will increase until market clearing occurs in one of the three markets.

If market clearing occurs first in the market for resources, then as w falls further the firm will operate in the RC case with the effective demands and supplies described above. From the comparative statics for RC, it is seen that as w falls the levels of labour bought and output sold will increase until market clearing occurs in one of these markets. If $\bar{Y} < F(\bar{N}^S, \bar{R}^S)$ then the first market to clear will be that for output; if $\bar{Y} > F(\bar{N}^S, \bar{R}^S)$, then the first market will be that for labour.

Assume that $\bar{Y} < F(\bar{N}^S, \bar{R}^S)$. As w falls further the firm will operate in the DRC region, with the firm on the long side of the markets for output and resource, and on the short side of the labour market. The effective demand for resource is thus R^{DCD}, which falls as w falls until the point where there is market clearing in the market for resource, after which the firm will operate in the DC region, on the long side of the market for output, but on the short side of the factor markets.

In the DC region the effective demand for labour is N^{SCD}, which increases as w falls until market clearing occurs in the labour market. At w below that point the firm operates in the DNC region, on the long side in the markets for labour and output, and on the short side in the market for resources, where it remains, since the effective demand for resources is inelastic with respect to the real wage.

Were it the case that $\bar{Y} > F(\bar{N}^S, \bar{R}^S)$, then the firm would always be on the short side in the market for output. Reducing w with the firm operating in the RC region will lead to market clearing in the market for labour, with the firm operating in the NRC for all lower w.

In the example above, the first market to clear when the firm was operating in the SC region was that for resource. At a higher level of (fixed) real resource price v the first market to clear might well be that for labour. As w falls further, the firm will operate in the NC region on the long side of the market for labour, on the short side of the other two markets. For inelastic labour supply the short sides of the market for output and the market for resource are inelastic with respect to the real wage, and so the firm operates in NC until zero real wage.

If $\bar{Y} < F(\bar{N}^S, \bar{R}^S)$ it is possible that the first market to clear while the firm is operating in SC is that for output, after which the firm operates in the DC region, and eventually, as w falls still further, the DNC region as described above.

Holding the real wage w constant and reducing the real resource price v leads to a symmetrical process with R replacing N and vice versa.

The slopes of the effective market-clearing loci can easily be determined by the analysis of Appendix B4 . With inelastic factor supplies, the lines Y—DR, N'—NR, and YN—DNR are horizontal as drawn, the lines Y—DN, R'—NR, and YR—DNR are vertical as drawn, and the lines O—DR and O—DN are straight lines as drawn. The other lines, although drawn as straight lines, are in general curved, and have been drawn with the relative slopes corresponding to a Cobb-Douglas production function with DRTS, as shown in Appendix B4.

In comparing Figures 3.3a, 3.3b, and 3.4 with Figures 2.6a, 2.6b, and 2.5, respectively, we can see the differences that explicit analysis of non-market-clearing trading leads to. The eight regions of Table 2.1 can be compared one-to-one with the eight regions of Table 3.1: the condition in each of the markets is identical in I and SC, II and RC, and so on. In particular, we note that in Figure 3.3a and Figure 2.6a, the region (DC or V) of generalized excess supply or Keynesian unemployment is much more extensive with non-market-clearing trading, and that region DNRC (or VIII) does not appear at all in this case. Comparing Figures 3.3b and 2.6b, we see that regions DRC, DNC, and DNRC (or VI, VII, and VIII) disappear completely in the case of non-market-clearing trading, and that region NRC (or IV), of generalized excess demand or repressed inflation, is much more extensive in this case. Regions DRC and DNC (or VI and VII) likewise do not appear in Figure 3.4 of non-market-clearing trading. These comparisons underline the error of using notional schedules in analyzing disequilibrium economies.

3.6.2. Comparative statics of the effective market-clearing loci.

Consider Figure 3.3a. A change in autonomous demand $\bar{Y}$ would not alter those sections of the resource- and labour-market-clearing loci corresponding to a sellers' market in the market for output (that is, those sections to the north and east of the output-market-clearing locus), since in that situation $\bar{Y}$ is not an argument of the effective schedules.

That is, as $\bar{Y}$ changed, the lines R—DR and N—DN would not shift. A change in $\bar{Y}$ would not affect Figure 3.3b, since none of the effective schedules depicted in that figure includes $\bar{Y}$ as an argument. In Figure 3.3a an increase in autonomous demand $\bar{Y}$ would be seen as a shift of the output-market-clearing locus Y—DR—DN—Y towards the origin, with the points DR and DN approaching each other until

$$\bar{Y} = F(\bar{N}^S, \bar{R}^S)$$

at which time Figure 3.3a would have become Figure 3.4. Further increase in $\bar{Y}$ would result in (unchanging) Figure 3.3b.

A change in resource supply $\bar{R}^S$ would shift the resource-market-clearing loci in the three figures, but would not alter those sections of the output- and labour-market-clearing loci corresponding to a buyers' market in the market for resource flow (that is, those sections to the east of the resource-market-clearing locus), since in that situation $\bar{R}^S$ is not an argument of the effective schedules. That is, as $\bar{R}^S$ changed, the only lines to shift would be Y—DR, N'—NR, and the resource-market-clearing locus itself. A decrease in $\bar{R}^S$ would be seen in Figure 3.3a as a shift of R—DR to the right, with point DR moving down the unshifting line DR—DN, O—DR rotating clockwise, and Y—DR shifting down parallel to its horizontal self. If the decrease continued, DR would approach DN until

$$\bar{Y} = F(\bar{N}^S, \bar{R}^S),$$

at which time Figure 3.3a would have become Figure 3.4. Further decrease of $\bar{R}^S$ would result in Figure 3.3b, with the resource-market-clearing locus R—NR—R' continuing its shift to the right, and NR moving down the unshifting labour-market-clearing locus NR—N. A change in labour supply $\bar{N}^S$ can be analysed symmetrically.

A change in the production technology can be analysed by examining its effect on the three markets. These effects are summarized in Table 3.6. In the case of neutral technical change, Appendix A3 shows that in the SC region the demands for both factor inputs increase, as does the supply of output. This is equivalent to a reduction in supply of both factor inputs, and a decrease in autonomous demand for output, in the market-clearing equations. Appendix B3 shows that in the DC region the demands for both factor inputs fall. This is equivalent to an increase in supply of the factor inputs. Appendix B5 shows that in the RC region the demand for labour and the supply of output increase, and that in the NC region the demand for resource flow and the supply of output increase. Thus effect of neutral technical change would be to shift all loci. Roughly, in Figure 3.3a, the output-market-clearing locus Y—DR—DN—Y would move away from the origin sufficiently rapidly that although section R—DR of the resource-market-clearing locus moved to the east, section O—DR of the locus could rotate in a counter-clockwise direction, and so that section N—DN of the labour-market-clearing locus would move to the north while section O—DN of the locus rotated in a clockwise direction. Arrows in Figure 3.5 indicate the movements. In Figure 3.3b, the resource-market-clearing locus R—NR—R' would move to the east and the labour-market-clearing locus N'—NR—N would move to the north until α was such that

$$(3.56) \qquad \bar{Y} = \alpha F(\bar{N}^S, \bar{R}^S), \qquad \alpha > 1,$$

Table 3.6. Summary of the effects of technical change.

Region	Neutral technical change $F^2(N, R) \equiv \alpha F(N, R)$		Resource-augmenting technical change $F^1(N, R) \equiv F(N, \alpha R)$	
	dN^D	dR^D	dN^D	dR^D
SC	+	+	+	+ α small - α large
RC	+	+	+	+ α small - α large
NC	+	+	+	+ α small - α large
DC	-	-	-	+ α small - α large

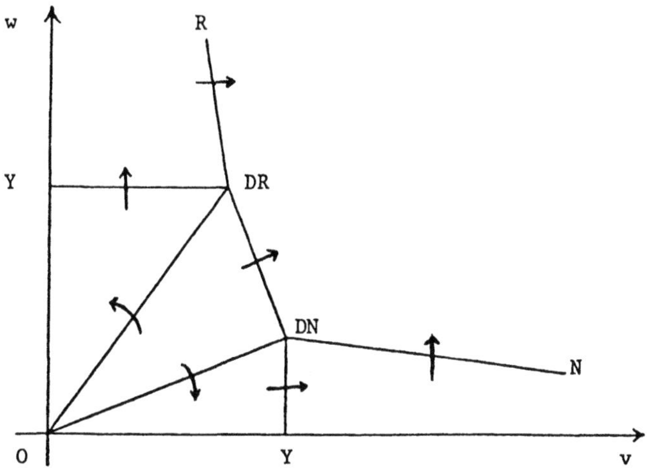

Figure 3.5a: The effect of neutral technical progress, $\bar{Y} < F(\bar{N}^S,\bar{R}^S)$.

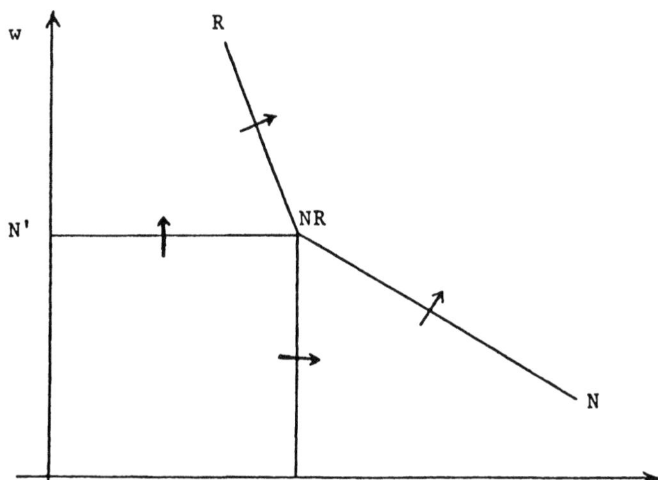

Figure 3.5b: The effect of neutral technical progress, $\bar{Y} > F(\bar{N}^S,\bar{R}^S)$.

at which time Figure 3.3b would have become Figure 3.4. Further increase in α would result in Figure 3.3a.

In the case of resource-augmenting technical change the movement of the resource-market-clearing locus varies with α, as derived in Appendices A3, B3, and B5, and as summarized in Table 3.6. Figures 3.6 and 3.7 show the movements of the loci, for small and large α, respectively.

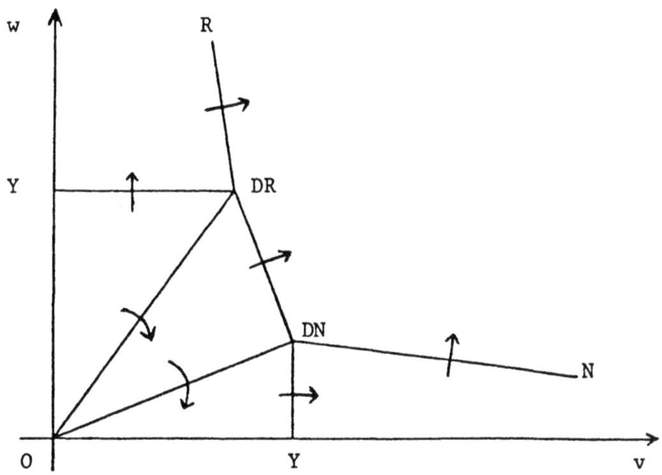

Figure 3.6a: The effect of resource-augmenting technical progress, $\bar{Y} < F(\bar{N}^S, \bar{R}^S)$, small α.

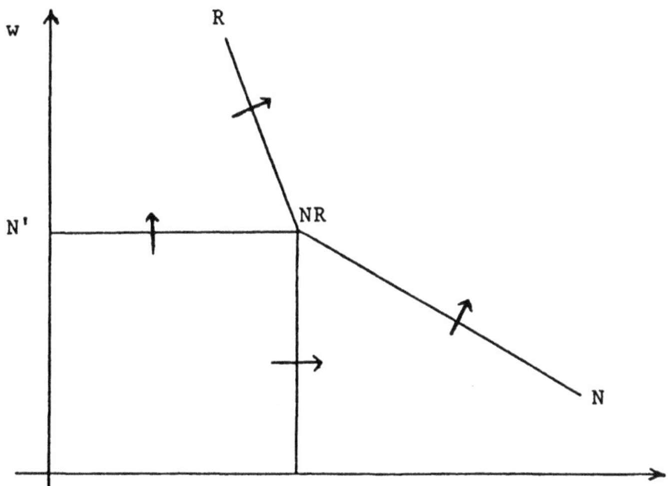

Figure 3.6b: The effect of resource-augmenting technical progress, $\bar{Y} > F(\bar{N}^S, \bar{R}^S)$, small α.

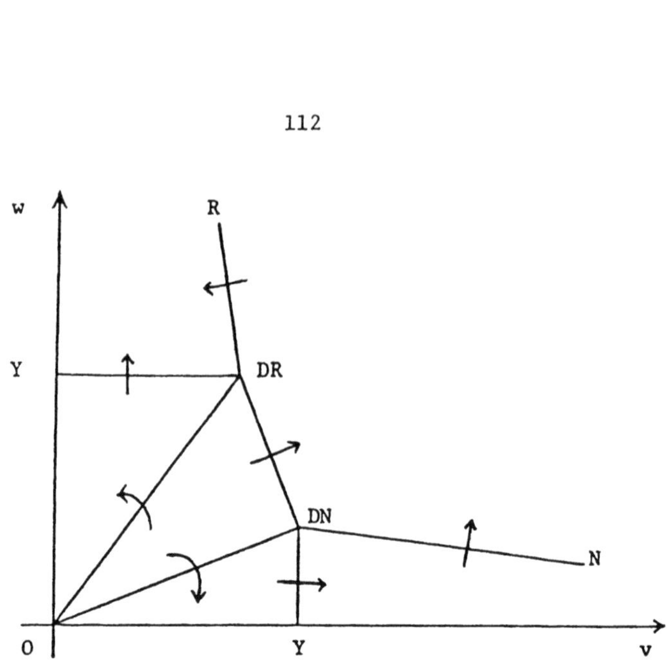

Figure 3.7 a: The effect of resource-augmenting technical progress, $\bar{Y} < F(\bar{N}^S, \bar{R}^S)$, large α.

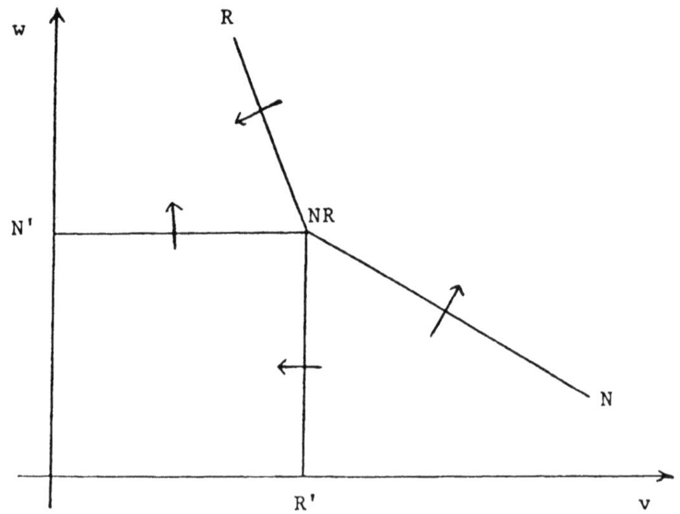

Figure 3.7 b: The effect of resource-augmenting technical progress, $\bar{Y} > F(\bar{N}^S, \bar{R}^S)$, large α.

CHAPTER IV: PRICE ADJUSTMENTS IN THE DISEQUILIBRIUM MODEL

4.1. Price determination.

Note that the previous discussion has implicitly assumed Jevon's Law of Indifference: that there is only one price in a competitive market. But, as Arrow (1959) remarks, there is no reason for profit- or utility-maximizing behaviour on the part of both sides of the market to lead to a unique price except in equilibrium or under conditions of perfect knowledge. In the situation where demand exceeds supply in the market for output, that is, when the representative firm (supplier of output) is on the short side of the market, then the firm can be thought of as a monopolist if each individual entrepreneur believes that raising his price will increase his profit. The process of adjustment, when each is attempting to raise his price, involves much uncertainty, and although there will be a broad tendency for prices to rise when demand exceeds supply, there can easily be considerable dispersion of prices among different sellers of the same good, and considerable variability over time in the rate of price change. Arrow characterizes the disequilibrium situation as a number of monopolists facing a number of monopsonists, and remarks on the value of information in consequence: entrepreneurs need to know the whole of the demand curve, not merely the current price.

Thus, in assuming a single price and (implicitly) perfect knowledge of the aggregate supply and demand on the part of individual buyers and sellers, this paper is not modelling Arrow's disequilibrium adjustment process. Rather, the process described in the preceding pages can be thought of as resulting from the existence of an auctioneer who calls a price, notes the resulting offers to buy and sell, and then announces the gross aggregate demand and supply at the price, after which trading will occur neither with market clearing nor with the possibility of recontracting. A Walrasian auctioneer would not be so passive--he would alter the price in an attempt to reduce the difference between aggregate supply and demand, and either no transactions would occur until a market-clearing price had been determined, or such transactions as did occur beforehand would be recontractable after the market-clearing price had been determined. If the process Arrow describes is more realistic than ours, ours in turn is more realistic than that of the Walrasian tâtonnement, and, moreover, with the addition below of finite rates of price adjustment, our formulation admits of the possibility, with reasonably simple analysis, of stable quasi-equilibria with price-inflation or -deflation and unemployment and demand-constrained recessions.

In Chapter III we analyzed the basic model in the case in which non-tatonnement trading is allowed. The model was presented in terms of its two state variables, the real wage w and the real resource flow price v. For any combination of w and v we were able to determine whether each market would be a sellers' market or a buyers' market or, on the boundaries, balanced. We were able to distinguish eight regions with imbalance in one or more markets and to plot the three loci of market claring, which bound these regions. The eight regions are categorized in Table 3.1, and plotted in Figures 3.3 and 3.4. Since we assume that quantities adjust infinitely faster than prices, for any point (w, v) we can determine the "disequilibrium" quantities supplied, demanded, and traded on each market. (We call such states disequilibria because in general they are not market-clearing, although Malinvaud (1977) maintains that they are "equilibria with rationing," since for a given price vector there is no tendency for quantities to change.)

Since in general for any (w, v) not all of the markets clear and there can be inadequate aggregate demand for output, these disequilibrium states have loosely been labelled as "Keynesian" (by Leijonhufvud (1968), Benassy (1973), and Drèze (1975), although both Grossman (1972) and Hahn (1976b) argue that this is not "what Keynes really meant." Whether or not it is we do not attempt to resolve here, but pass on, noting that the approach yields a rich theoretical harvest and new insights into disequilibrium processes in market economies.

In this chapter we allow the prices to adjust in response to the imbalances of the three markets. We shall show the existence of "quasi-equilibria," points at which the real prices are constant although in general the three markets do not clear and the money prices continue to adjust. For any combination of values of exogenous variables we shall see that there exists only one quasi-equilibrium, although the disequilibrium region in which it occurs varies with the exogenous variables' values. Among the exogenous variables which affect the position of the quasi-equilibrium, we shall see that the speeds of adjustment of money prices are important. And we shall see that the various formulations of real price adjustment also affect the existence and stability of the quasi-equilibrium. Although in Chapter V the supply of resource flow will no longer be inelastic, the positions of the quasi-equilibria as derived in this chapter will be important, since at quasi-equilibrium the expected rate of change of real resource flow price e* and hence the supply of resource flow $R^S(e*)$ will be constant. (See sections 5.2 and 5.3.)

In the basic model of Chapter II, given the assumption of recontracting (or the equivalent assumption of no non-market-clearing exchange), the dynamic analysis of section 2.4.2 was concerned simply with the convergence of the prices to their general-market-clearing levels during the tatonnement process. By dropping the assumption of no non-market-clearing trading, and recognizing that most contracts rule out the possibility of recontracting, we are able to examine the dynamic behaviour of the actual adjustment of the economy, implicitly including adjustments in quantities, in contrast to the "virtual" adjustments of Chapter II.

In section 2.4.2 we proposed the concept of the mythical Walrasian auctioneer, whose task it was to revise the nominal price on each market to reduce the difference between (notional) excess demand and supply on that market. The end result, if this process converged, would be the general-market-clearing price vector. We assumed that these price-setting agents knew the signs of the elasticities of the notional demand and supply functions, and that they would increase or decrease the nominal price depending on whether they observed excess demand or excess supply for the good. The discussion of Chapter III had nothing to say about such dynamic behaviour.

In order to study the adjustment of the economy with flexible prices, we need a theory of price determination. Ideally such a theory would embody optimizing behaviour on the part of the economic actors involved: the commonly used Walrasian excess demand hypothesis, which states that in competitive markets prices will rise when there is excess demand and fall when there is excess supply, has not been derived as the optimizing response of economic units to changing data. In work on a profit-maximizing firm in uncertain markets, able to adjust the price of its output, the wage offered, and the number of employees through virtue of its quasi-monopolistic position, Iwai (1974) was able to derive price adjustment relationships similar to the Walrasian excess demand hypothesis. VanOrder (1976), with the similar assumption of a price- and wage-setting firm, derived a similar relationship. But we have assumed that all actors are price-takers (although this assumption will be relaxed in examining adjustments in region DNRC below).

The economic motivation for a model of price determination in a competitive market remains a topic of continuing discussion (Koopmans (1957), Arrow (1959), Negishi (1962), Hansen (1970), and Hahn (1976b)), but for the moment we shall continue to use the familiar Walrasian excess demand hypothesis of Chapter II, adapted for the non-market-clearing trading model of Chapter III. As Patinkin (1965) pointed out, with non-market-clearing trading the consequent rationing of the long side in any market will result in "spillover" effects: demand and supply functions will depend not only on price signals, but also on quantity signals. This realization led Clower (1965) to formulate effective, rather than notional, supply and demand schedules. A consequence of these spill-overs is that the movements of a price on one market could be influenced by imbalances on other markets. Effective schedules already embody the spillover effects of quantity-rationing, and will consequently be used instead of notional schedules in the formulation below: competitive prices will rise when there is excess <u>effective</u> demand and fall when there is excess <u>effective</u> supply.

The general form of the Walrasian excess demand hypothesis can be written

$$(4.1) \qquad \frac{1}{p}\frac{dp}{dt} = f(q^D - q^S) = f(q^X), \qquad f' > 0,$$

where dp/dt is the time rate of change of nominal price, and q^X is the excess effective demand on the market. If f(0) is non-zero, there is said to be autonomous price change. We shall formulate a particular relationship for each of the three nominal (or money) prices in the model: P, W, and V. Given the definitions of the real prices,

$$(4.2) \qquad w \equiv W/P \quad \text{and} \quad v \equiv V/P,$$

this will result in two equations for the proportional rates of change of the real prices,

$$(4.3) \qquad g \equiv \dot{v}/v = \dot{V}/V - \dot{P}/P$$

$$h \equiv \dot{w}/w = \dot{W}/W - \dot{P}/P.$$

With no autonomous price change, zero excess effective demand in the three markets is a sufficient condition for constant real wage w and constant real resource flow price v. (This situation of triple market clearing is depicted as region DNRC in Figure 3.4, reproduced as Figure 4.1) But with exogenous labour supply $\bar{N}^S$, resource flow supply $\bar{R}^S$, and autonomous demand for output $\bar{Y}$, in general there is no point (w, v) at which all three markets clear. This, it was argued at some length in section 2.3.1, would require

$$\bar{Y} = F(\bar{N}^S, \bar{R}^S),$$

but in general autonomous demand will not equal full-employment output, and the market-clearing loci can be depicted either by Figure 3.3a or by Figure 3.3b, reproduced as Figures 4.2a and 4.2b.

A necessary condition for constant real wage w is seen to be that the nominal wage W and the nominal price of output P (taken as a proxy for the overall price level) change at equal proportional rates, $\dot{W}/W = \dot{P}/P$. Similarly, a necessary condition for constant real resource flow price v is seen to be that the nominal resource flow price V and the nominal price of output P change at equal proportional rates, $\dot{V}/V = \dot{P}/P$. Thus we see that we can define a state in which real prices are constant, although markets do not clear and there exist non-zero excess effective demands in the economy. This concept of equilibrium is Hansen's (1951) "quasi-equilibrium," which has been used in several recent studies: Solow and Stiglitz (1968), Benassy (1973), Korliras (1975), Barro and Grossman (1976), and Varian (1976).

The concept of the quasi-equilibrium includes as a special case the Walrasian general-market-clearing equilibrium. It allows examination of states of the economy with non-market-clearing trading. In this chapter we examine the existence, uniqueness, and stability of possible quasi-equilibria, using two formulations of the Walrasian excess demand hypothesis of price change. We also examine the behaviour of the firms in region DNRC, where all markets clear. We argue that, despite this general market clearing, nominal prices may adjust, due to competition between identical, price-setting firms, with resulting changes in the real prices w and v.

4.1.1. <u>SF: The simple formulation of price adjustment</u>.

In this section we adapt the formulation of section 2.4.2 to take account of the spillover effects resulting from non-market-clearing trading. For the labour market, the simple formulation (SF) case can be written

$$\dot{W}/W = \lambda_W N^X, \tag{4.4}$$

where the positive constant λ_W is the speed of money wage adjustment. This equation states that the proportional rate of change of the nominal (or money) wage is an increasing function of the excess effective demand on the labour market.

For the resource flow market, the SF case can be written

$$\dot{V}/V = \lambda_V R^X, \tag{4.5}$$

where the positive constant λ_V is the speed of adjustment of the money price of resource flow. The equation states that the proportional rate of change of the nominal (or money) price of resource flow is an increasing function of the excess effective demand on the market for resource flow.

For the output market the SF case can be written

$$\dot{P}/P = \lambda_P Y^X, \tag{4.6}$$

where the positive constant λ_P is the speed of adjustment of the money price of output, a proxy for the overall price level. This formulation models "demand-pull" price inflation, with no autonomous component, and states that the proportional rate of change of the nominal (or money) price of output is an increasing function of the excess effective demand on the market for output. An alternative formulation in section 4.1.2 below includes a "cost-push" inflation component.

The proportional rates of change of the real price of resource flow and the real wage are given by

$$(4.7) \qquad g \equiv \dot{v}/v = \lambda_V R^X - \lambda_P Y^X$$

$$h \equiv \dot{w}/w = \lambda_W N^X - \lambda_P Y^X.$$

Since the excess effective demands R^X, N^X, and Y^X, as summarized in Table 4.1, from Table 3.4, are functions of w, v, and the exogenous variables $\bar{N}^S$, $\bar{R}^S$, and $\bar{Y}$, it is possible to plot phase diagrams for the economy, showing the constant-w and constant-v loci, and the tendencies for the two real prices to change in the different areas of the plane.

Consider Figure 4.1, which is similar to Figure 3.4 and shows the three market-clearing loci when the autonomous demand for output $\bar{Y}$ equals the full-employment output. In each region the signs of the three excess effective demands are shown as a vector:

$$(Y^X, \quad N^X, \quad R^X).$$

Figures 4.2a and 4.2b, similar to Figures 3.3a and 3.3b, illustrate the economy when the autonomous demand $\bar{Y}$ is less than, and greater than, respectively, the full-employment output. The signs of the three nominal price changes are summarized in Table 4.2, which also includes the signs of the rates of change of the two real prices in each region. Table 4.2 is seen to be similar to Table 2.1, except for the final region, DNRC, which is one of general market clearing.

Table 4.1. Excess effective demands in the eight regions.

Region	Output, Y^X	Labour, N^X	Resource flow, R^X
SC	$s(\bar{Y} - Y^{SCS}) > 0$	$N^{SCD} - \bar{N}^S < 0$	$R^{SCD} - \bar{R}^S < 0$
RC	$s(\bar{Y} - Y^{RCS}) > 0$	$N^{RCD} - N^S < 0$	$R^{SCD} - \bar{R}^S > 0$
NC	$s(\bar{Y} - Y^{NCS}) > 0$	$N^{SCD} - \bar{N}^S > 0$	$R^{NCD} - \bar{R}^S < 0$
NRC	$s(\bar{Y} - Y^{NRCS}) > 0$	$N^{RCD} - \bar{N}^S > 0$	$R^{NCD} - \bar{R}^S > 0$
DC	$\bar{Y} - Y^{SCS} < 0$	$N^{DCD} - \bar{N}^S < 0$	$R^{DCD} - \bar{R}^S < 0$
DRC	$\bar{Y} - Y^{RCS} < 0$	$N^{DRCD} - \bar{N}^S < 0$	$R^{DCD} - \bar{R}^S > 0$
DNC	$\bar{Y} - Y^{NCS} < 0$	$N^{DCD} - \bar{N}^S > 0$	$R^{DNCD} - \bar{R}^S < 0$
DNRC	$\bar{Y} - Y^{NRCS} = 0$	$N^{DRCD} - \bar{N}^S = 0$	$R^{DNCD} - \bar{R}^S = 0$

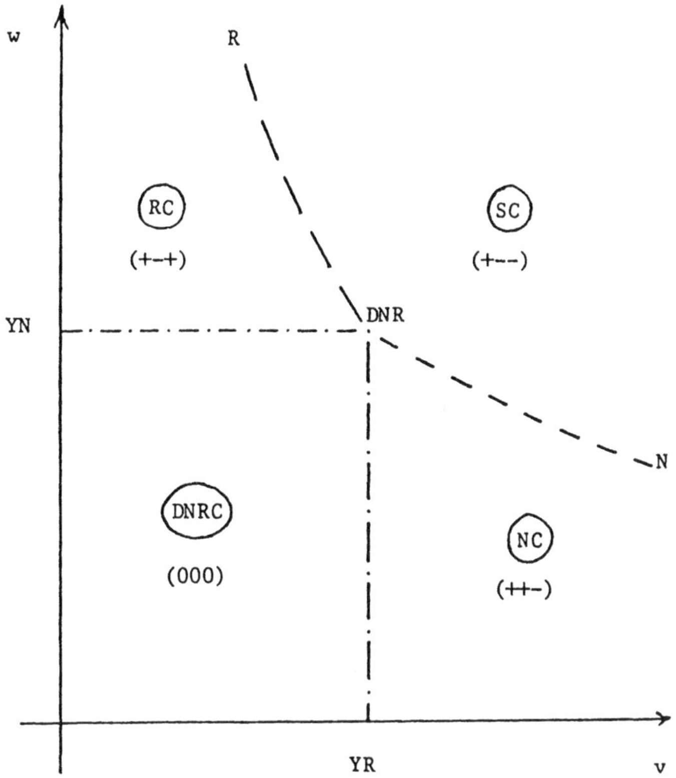

Figure 4.1: The effective market-clearing loci, $\bar{Y} = F(\bar{N}^S, \bar{R}^S)$.

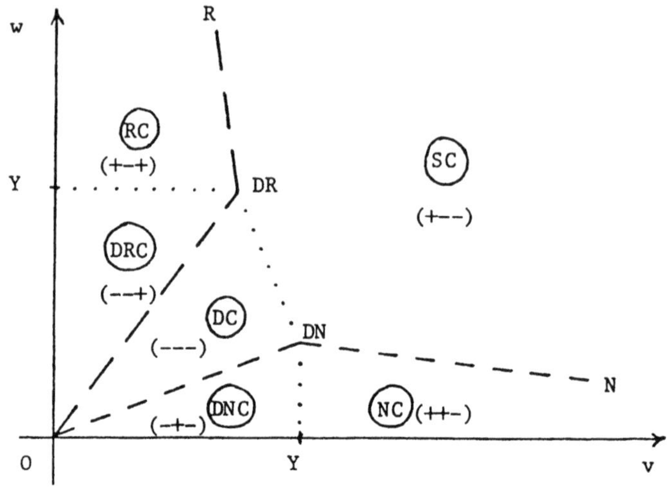

Figure 4.2a: The effective market-clearing loci, $\bar{Y} < F(\bar{N}^S, \bar{R}^S)$.

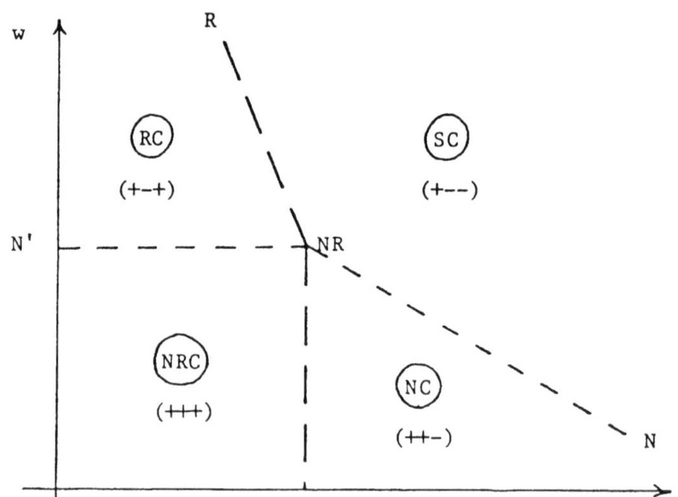

Figure 4.2b: The effective market-clearing loci, $\bar{Y} > F(\bar{N}^S, \bar{R}^S)$.

Table 4.2. The signs of the price changes with the simple formulation.

Region	$\dot{P}/P$, Y^X	$\dot{W}/W$, N^X	$\dot{V}/V$, R^X	$h \equiv \dot{w}/w$	$g \equiv \dot{v}/v$
SC	+	-	-	-	-
RC	+	-	+	-	±
NC	+	+	-	±	-
NRC	+	+	+	±	±
DC	-	-	-	±	±
DRC	-	-	+	±	+
DNC	-	+	-	+	±
DNRC	0*	0*	0*	0*	0*

*Note: We argue in the text that although all markets clear in region DNRC, nominal prices (and real prices) might still change in response to firms' competition.

Table 4.3 summarizes the signs of the elasticities of the excess effective demands. It is derived from Tables 4.1 and 3.3. The signs of the elasticities of the excess effective demand in the output market were derived using the discussion in section 3.4.1, and are opposite from the signs of the elasticities of the supply of output as noted in that discussion. From Table 4.3 we can note that all excess effective demands are inelastic in region DNRC, the region of general market clearing.

Points of quasi-equilibrium of the system occur when $\dot{w} = 0$ and $\dot{v} = 0$, that is, at the points of intersection of the constant-w and constant-v loci, both of which can be derived from the pair of equations above. Consider Figure 4.3, in which the autonomous demand for output $\bar{Y}$ equals the full-employment output, $F(\bar{N}^S, \bar{R}^S)$. We shall argue below that the constant-v locus passes through point DR in Figure 4.4a, where the markets for resource flow and output both clear. Figure 4.3 is a limiting case of Figure 4.4a, and so we argue that the constant-v locus passes through point DNR, where all three markets clear. In region RC, $\dot{P}/P$ is the same sign as $\dot{V}/V$, so that $\dot{v}/v$ will be positive, zero, or negative depending on the relative sizes of the two. When they are equal, $\dot{v}$ is zero. The figure shows the constant-v locus passing through region RC to point DNR. By similar argument, the constant-w locus can be shown to pass through the NC region to point DNR. Point DNR is a point of equilibrium, with all three markets clearing, and constant nominal prices W, V, and P, as well as constant real prices w and v. But in Appendix B1 we argue that it is not the only point of general market clearing in Figure 4.3, since region DNRC, if it exists (that is,

Table 4.3. The signs of the elasticities of the excess effective demands.

Region	Y_w^X	Y_v^X	N_w^X	N_v^X	R_w^X	R_v^X
SC	+	+	-	-	-	-
RC	+	0	-	0	-	-
NC	0	+	-	-	0	-
NRC	0	0	-	0	0	-
DC	+	+	-	+	+	-
DRC	+	0	0	0	+	-
DNC	0	+	-	+	0	0
DNRC	0	0	0	0	0	0

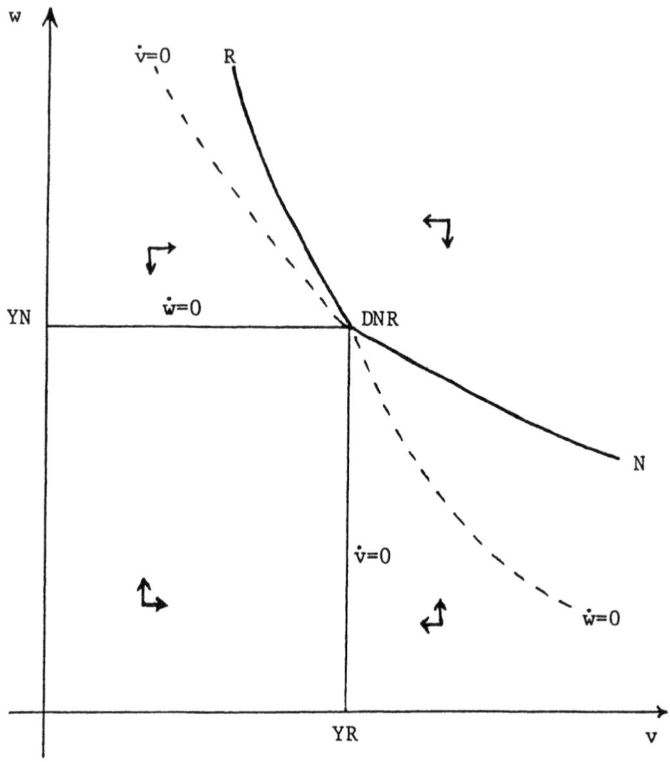

Figure 4.3: Phase diagram, simple formulation, $\bar{Y} = F(\bar{N}^S, \bar{R}^S)$.

if $\bar{Y} = F(\bar{N}^S, \bar{R}^S)$) is a region of general market clearing. Thus, following the Walrasian excess demand hypothesis, we should expect the nominal prices, and hence the real prices, to be constant throughout the region DNRC.

In Appendix B1, equations (B40), we derived the market-clearing conditions for region DNRC:

$$(4.8) \qquad Y^D = \bar{Y} = F(\bar{N}^S, \bar{R}^S) = Y^{NRCS},$$

$$N^{DRCD}(\bar{Y}, \bar{R}^S) = \bar{N}^S,$$

$$R^{DNCD}(\bar{Y}, \bar{N}^S) = \bar{R}^S,$$

which state that the effective demand in each factor input market, derived with the Benassy assumption of ignoring the possibility of being constrained on that market, is just equal to the supply of that input, and that the resulting effective supply of output is just equal to the effective demand for output. In deriving these conditions, we used the convenience, introduced in section 2.1, of the representative firm, whose behaviour, except for its scale, is identical with the aggregate of such units.

Let us look behind the veil of the "representative" firm. For region DNRC we assume instead a large number of identical, price-setting firms, whose aggregate behaviour on the three markets is described by equations (4.8) above. Thus we have assumed an implicit rationing scheme whereby autonomous demand for output $\bar{Y}$, labour supply $\bar{N}^S$, and supply of resources $\bar{R}^S$ are allocated between firms. Let us now ask what the result in aggregate would be of all firms trying to improve their allocations by offering to hire labour at a higher money wage W, by offering to buy resource flow at a higher money price V, and by offering for sale output at a lower money price P. In aggregate, equations (4.8) would still hold, and competition between the identical firms would stymie the attempt of each firm to increase its allocation of scarce supplies of labour and resource flow, and scarce demand for output. But a result of this competition between identical, price-setting firms would be increases of the real wage w and the real resource price v. Firms could afford to offer higher W and V, and lower P because, as shown in Appendix B1, in region DNRC they operate with the marginal product of each factor input greater than its real price.

In Figure 4.3 these movements would be seen as a movement to the north and east in region DNRC, towards the triple-market-clearing point DNR, as indicated by the small arrows. Above line YN—DNR, in region RC, there is excess effective demand in the market for output and excess effective supply in the labour market: firms can hire as much labour, and sell as much output, as they wish, given the supply of resource flow, $\bar{R}^S$. Thus in this region identical, price-setting firms would tend to increase V to increase their individual allocations of the scarce supply of resource flow $\bar{R}^S$, and profit maximizing would tend to result in lower money wages W and higher money price of output P, with consequent rising real resource flow price v and falling real wage w, as deduced from the Walrasian excess demand hypothesis. Thus line YN—DNR can be thought of as an extension of the constant-w locus, N—DNR. A similar argument for region NC, to the right of line YR—DNR, would conclude that the Walrasian excess demand hypothesis modelled the consequences of identical, price-setting firms in that region: higher P and W, lower V, with resulting falling v and rising w, as indicated in Figure 4.3. Thus line YR—DNR can be thought of as an extension of the constant-v locus, R—DNR.

In all other regions of Figures 4.3, 4.4a, and 4.4b, similar arguments would show that the Walrasian excess demand hypothesis accurately models the consequences of nominal price competition between identical, price-setting firms: the nominal price will rise in a buyers market and fall in a sellers' market. These arguments agree with the more rigorous work of Iwai (1974) and VanOrder (1976) mentioned in section 4.1 above.

Consider Figure 4.4a, with $\bar{Y} < F(\bar{N}^S, \bar{R}^S)$. The constant-v locus cannot pass through regions in which the excess effective demands for resource flow and output are of opposite sign, which rules out regions SC, NC, and DRC. It must pass through point DR, where both the money price of resource flow and the money price of output are constant since the markets for resource flow and output both clear at this point. Thus the constant-v locus must pass through regions RC, DC, and DNC as shown. By similar argument, the constant-w locus does not pass through regions SC, RC, and DNC, but does pass through regions NC, DC, and DRC, and through point DN, as shown. The slopes of the two constant real price loci depend on the relative sizes of λ_W, λ_V, and λ_P, as shown in Appendix C1. They can only intersect in the DC region of Keynesian unemployment, resulting in quasi-equilibrium Q_{SF}^{DC}, where both real prices are constant, although the three money prices are falling.

Consider Figure 4.4b, with $\bar{Y} > F(\bar{N}^S, \bar{R}^S)$. The constant-v locus cannot pass through regions SC or NC, nor through point NR, since at this point the money price of output is increasing, but the money wage and the money price of resource flow are both constant, with the two factor input markets clearing. At point NR, then, both real prices w and v are falling. The constant-v locus passes through regions RC and NRC as shown. By similar arguments the constant-w locus passes through regions NC and NRC as shown. The two loci can only intersect in the NRC region of repressed inflation, resulting in quasi-equilibrium Q_{SF}^{NRC}, where both real prices are constant, although the three money prices are rising. Since, as shown in Appendix C1, the constant-v locus is vertical and the constant-w locus is horizontal in region NRC, Q_{SF}^{NRC} is unique.

It is readily apparent from Table 4.3 that to the left of the constant-v locus, v is increasing, and to the right, decreasing, and that below the constant-w locus, w is increasing, and above, decreasing. The small arrows in the phase diagrams indicate these tendencies. In section 4.2.1 below, we consider the stability of these quasi-equilibria, and in section 4.3.1, the effect on them of changes in the exogenous variables.

We can easily see that the position of the loci, and hence the position of the quasi-equilibrium, is a function of the relative speeds of adjustment, λ_V, λ_W, and λ_P. As the ratio λ_V/λ_P tends to infinity, the constant-v locus approaches the locus of resource market clearing (R—DR—O in Figure 4.4a, R—NR—R' in Figure 4.4b). As the ratio λ_V/λ_P tends to zero, the constant-v locus approaches the locus of output market clearing, Y—DR—Y, in Figure 4.4a, or the w-axis in Figure 4.4b. Symmetrical shifts of the constant-w locus occur as the ratio λ_W/λ_P varies.

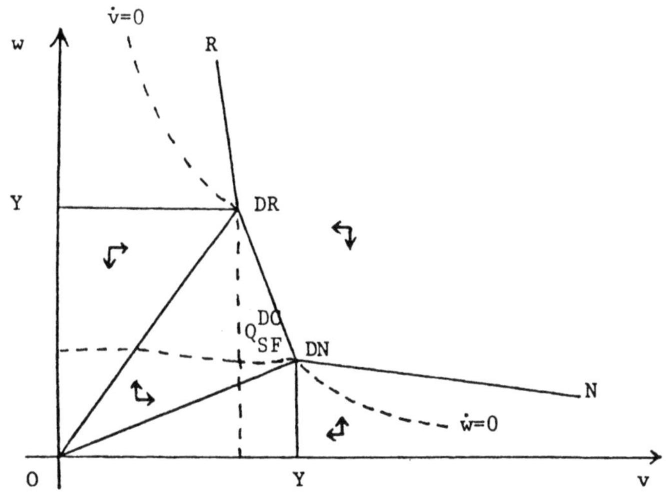

Figure 4.4a: Phase diagram, simple formulation, $\bar{Y} < F(\bar{N}^S, \bar{R}^S)$.

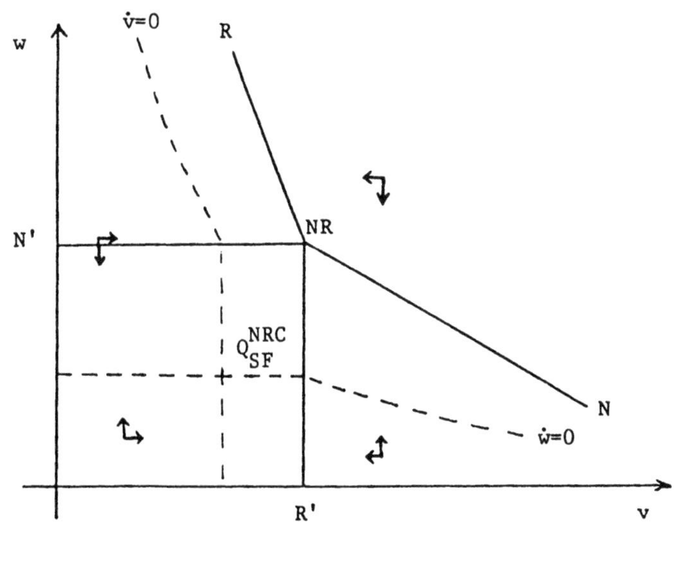

Figure 4.4b: Phase diagram, simple formulation, $\bar{Y} > F(\bar{N}^S, \bar{R}^S)$.

4.1.2. AF: The alternative formulation of price adjustment.

In this section we examine a price adjustment formulation which includes elements of "cost-push" price inflation, and which can model "money illusion" in the factor input markets. It was suggested by the formulation in Solow and Stiglitz (1968). For the labour market, the alternative formulation (AF) case can be written

$$\text{(4.9)} \qquad \dot{W}/W = \lambda_W N^X + m \cdot \dot{P}/P, \qquad 0 \le m \le 1.$$

This states that the proportional rate of change of money wage is the sum of a function decreasing in unemployment and of a fraction of the proportional rate of change of the money price of output. The fraction m is a decreasing function of "money illusion" in the labour market: if m is nearly one, the wage bargain is struck in very nearly real terms, and the expected rate of change of prices is, *ex post*, very nearly fulfilled, where

$$\text{(4.10)} \qquad h \equiv \dot{w}/w = \lambda_W N^X - (1-m)\, \dot{P}/P.$$

If m is closer to zero, then there is more money illusion in the labour market: a stronger tendency to respond to changes in money terms as though they represented changes in real terms, that is, there is a stronger tendency to underestimate changes in the price level.

For the resource flow market, the alternative formulation case can be written

$$\text{(4.11)} \qquad \dot{V}/V = \lambda_V R^X + n \cdot \dot{P}/P, \qquad 0 \le n \le 1.$$

This states that the proportional rate of change of money resource price is the sum of a function increasing in excess effective demand on the resource flow market and of a fraction of the rate of price inflation. The fraction n, analogous with m in the labour market, is a decreasing function of money illusion in the resource flow market, where

$$(4.12) \qquad g \equiv \dot{v}/v = \lambda_V R^X - (1-n)\,\dot{P}/P.$$

For the output market, the alternative formulation can be written

$$(4.13) \qquad \dot{P}/P = \lambda_P Y^X + j\cdot\dot{W}/W + k\cdot\dot{V}/V, \qquad 0 \leq (j+k) < 1;\ j,k > 0.$$

This states that the money price of output is partly cost-determined, the rate of change of money price of output (the rate of price inflation) depending on the excess effective demand in the output market, on the rate of change of labour cost per unit of output (or of unit labour cost at some standard rate of output), and on the rate of change of resource cost per unit of output (or of unit resource cost at the standard rate of output). In a short-run model such as this, productivity (of either input factor) can be treated as constant at the standard output, in which case fluctuations in standard unit factor input cost are proportional to fluctuations in the money price of the factor. Thus the third equation is derived, with a reasonable limit on the effect of unit factor input cost.

The reader will note that this formulation is very similar to the Phelps-Friedman variant of the Phillips curve (Phelps (1968), Friedman (1968)) which can be written in terms of three markets as

$$(4.14) \qquad \begin{aligned} \dot{W}/W &= \lambda_W N^X + p \\ \dot{V}/V &= \lambda_V R^X + p \\ \dot{P}/P &= \lambda_P Y^X + j.\dot{W}/W + k.\dot{V}/V, \qquad j+k=1,\ j,k > 0 \\ \dot{p} &= \lambda.(\dot{P}/P - p), \qquad \lambda > 0 \end{aligned}$$

where $p \equiv \dot{P}^e/P$ is the expected rate of price level change which is based on adaptive expectations so that there is a short-run unemployment-inflation tradeoff.

In the Phelps-Friedman formulation (4.14), the rate of change of money wage can be thought of as the outcome of the following process: the labour market behaves as if it were competitive and adjusts real wages to remove excess (effective) demand. But since money wages, rather than real wages, are set in the labour market, the money wages will be set in accordance with excess demand and the expected rate of price change. The coefficient on the expected rate of price change is predicted to be unity from the assumption that the excess demand for labour is homogeneous of degree zero in the money and price level. A similar process in the resource flow market leads to the equation for the rate of change of money resource flow price in equations (4.14).

Subtracting the actual rate of change of prices, $\dot{P}/P$, from both sides of the first two equations (4.14), we obtain

$$(4.15) \qquad \begin{aligned} \dot{w}/w &= \lambda_W N^X - (\dot{P}/P - p) \\ \dot{v}/v &= \lambda_V R^X - (\dot{P}/P - p). \end{aligned}$$

It can be seen that the closer the expected rate of change of price level to the actual, the closer the factor price bargain to real terms. Note that in the Phelps-Friedman model, if $p > \dot{P}/P$, the rate of change of real factor prices will be greater than the component from the excess demand on the factor market, a situation which the alternative formulation cannot admit, since both m and n are less than one in equations (4.9) and (4.11).

Substituting the adaptive expectations equation from equations (4.14) into equations (4.15), we obtain

$$\text{(4.16)} \qquad \dot{w}/w = \lambda_W N^X - \dot{p}/\lambda$$

$$\dot{v}/v = \lambda_V R^X - \dot{p}/\lambda.$$

It can be seen that as the speed of adjustment of price level expectations increases, the factor price bargain is struck closer to real terms.

In comparing the Phelps-Friedman formulation with our alternative formulation of equations (4.17), we see that the case of no money illusion in the former ($m = 1$ in the latter) corresponds to infinite λ: expectations adjust so quickly that the bargain is struck in real terms. Complete money illusion ($m = 0$ in the alternative formulation) corresponds to zero λ in the Phelps-Friedman formulation: expectations are static, the expected rate of change of price level is constant and not affected by the experience of the actual rate. It is possible, in the Phelps-Friedman model, that the (unchanging) expectation in this case is correct at any instant, resulting in the wage bargain's being struck in real terms, but in general expectations are not fulfilled, and there is a tendency to respond to changes as though they represented changes in real terms. The treatment for the resource flow market and the parameter n is similar.

Thus we can see two advantages of our alternative formulation over the Phelps-Friedman formulation: firstly, we can easily set $m \neq n$ to allow for different degrees of money illusion (different speeds of adjustment of price level expectations) on the two factor input markets, which may more accurately model differences between the two groups of sellers: resource owners and workers. Secondly, our formulation has the technical advantage of allowing us to dispense with the additional state variable p, the expected rate of price level change, and thus have one less degree of freedom to deal with in the model.

Manipulation of the three equations (4.9), (4.11) and (4.13) results in the elimination of $\dot{P}/P$, to give the proportional rates of change of the real price of resource flow and the real wage:

$$(4.17) \qquad g \equiv \dot{v}/v = \lambda_V^V R^X - \lambda_P^V Y^X - \lambda_W^V N^X$$

$$h \equiv \dot{w}/w = \lambda_W^W N^X - \lambda_P^W Y^X - \lambda_V^W R^X$$

where the six constants are functions of λ_W, λ_V, λ_P, j, k, m, and n, as stated in Appendix C2.

Table 4.4 summarizes what we can say about the signs of the rates of change of the three nominal prices $\dot{P}$, $\dot{W}$, and $\dot{V}$, and the two real prices $\dot{w}$ and $\dot{v}$. Since the excess effective demands R^X, N^X, and Y^X are functions of the exogenous variables and w and v, it is possible to plot phase diagrams for the economy, showing the constant-w and the constant-v loci.

Consider Figure 4.5, which shows the three market-clearing loci when the autonomous demand for output $\bar{Y}$ is equal to the full-employment output. The constant-v locus will pass through point DNR, where all three markets clear. In Appendix B1 we argue that if region DNRC exists, that is, if $\bar{Y} = F(\bar{N}^S, \bar{R}^S)$, then there is triple market clearing throughout the region. By arguments similar to those in section 4.1.1 above, real prices w and v will rise in region DNRC as a result of competition

Table 4.4. The signs of the price changes with the alternative formulation.

Region	$\dot{P}/P$	$\dot{W}/W$	$\dot{V}/V$	$h \equiv \dot{w}/w$	$g \equiv \dot{v}/v$
SC	±	±	±	±	±
RC	±	±	±	-	±
NC	±	±	±	±	-
NRC	+	+	+	±	±
DC	-	-	-	±	±
DRC	±	±	±	±	+
DNC	±	±	±	+	±
DNRC	-	+	+	+	+

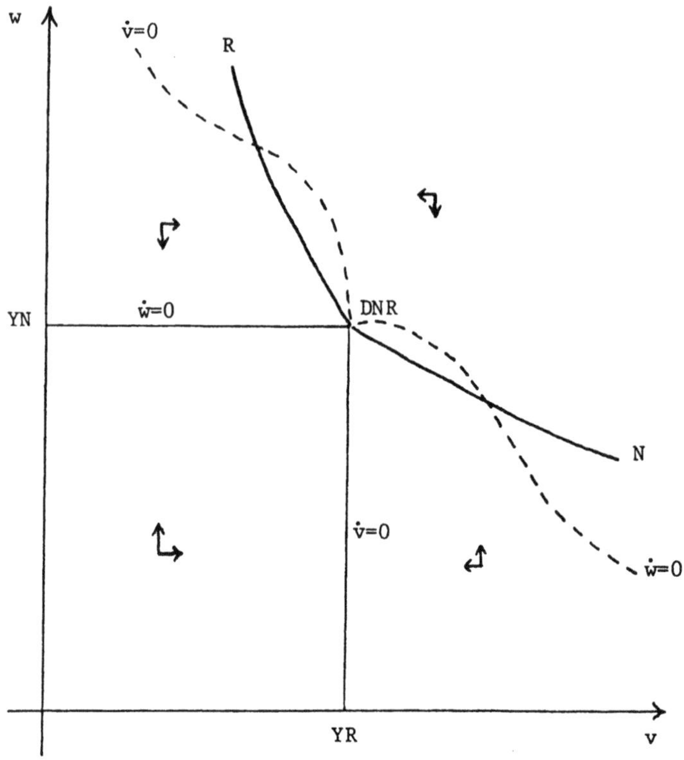

Figure 4.5: Phase diagram, alternative formulation, $\bar{Y} = F(\bar{N}^S, \bar{R}^S)$.

between price-setting firms. Along the line R—DNR, the resource flow market clears and R^X is zero. But g can be positive, zero, or negative, depending on the relative magnitudes of the other two components. The constant-v locus in Figure 4.5 has been plotted to pass through both region RC and region SC. By similar argument, the constant-w locus will pass through point DNR, and the NC and SC regions depending on the relative magnitude of the two parameters. Point DNR is a point of general market clearing, with nominal, as well as real, prices constant. Elsewhere, the small arrows indicate the tendency of movement of the state variables w and v, in response to unbalanced markets, or in region DNRC price-setting competition.

Consider Figure 4.6a, with $\bar{Y} < F(\bar{N}^S, \bar{R}^S)$. Examination of the excess effective demands in the various regions, summarized in Table 4.4, shows that the constant-v locus passes through regions RC, SC, DC, and DNC. Similarly, the constant-w locus passes through regions NC, SC, DC, and DRC. At point DR, v is rising and w is falling since N^X is negative while R^X and Y^X are zero. At point DN, v is falling and w is rising since R^X is negative while N^X and Y^X are zero. In Figure 4.6a the two loci are shown intersecting at Q^{SC}_{AF}, a point of quasi-equilibrium in the region of classical unemployment. The point of intersection could also occur in region DC, the region of Keynesian unemployment with falling money prices, quasi-equilibrium Q^{DC}_{AF} (not shown), or even on the output-market-clearing locus between points DN and DR.

Consider Figure 4.6b, with $\bar{Y} > F(\bar{N}^S, \bar{R}^S)$. At point NR, N^X and R^X are zero, while Y^X is positive. This means that at NR both v and w are falling, and the constant-v and constant-w loci pass to the west and

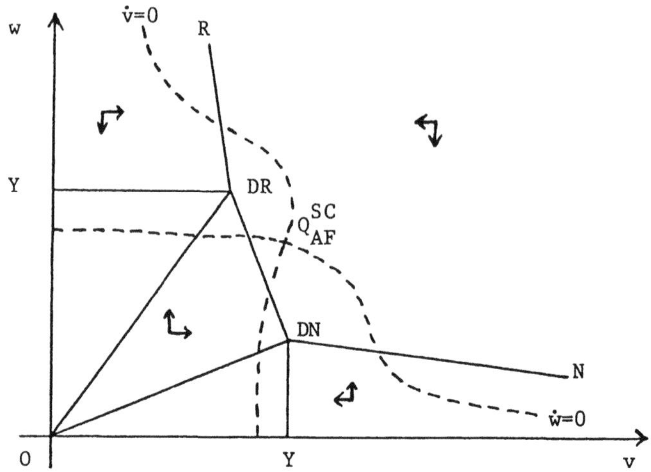

Figure 4.6a: Phase diagram, alternative formulation, $\bar{Y} < F(\bar{N}^S, \bar{R}^S)$.

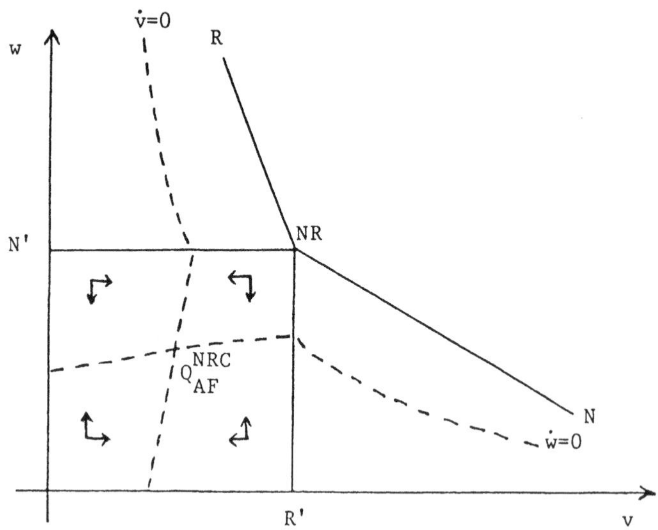

Figure 4.6b: Phase diagram, alternative formulation, $\bar{Y} > F(\bar{N}^S, \bar{R}^S)$.

south of the point, given the elasticities of the excess demands of Table 4.3. The two loci can thus intersect only in region NRC, the region of repressed inflation and rising money prices, to result in the quasi-equilibrium Q_{AF}^{NRC}, as shown. In Appendix C2 we show that the slopes of the two loci in region NRC are positive.

It is readily apparent from Table 4.3 that to the left of the constant-v locus v is increasing, and to the right, decreasing, and that below the constant-w locus w is increasing, and above, decreasing. The small arrows in the phase diagrams indicate these tendencies. In section 4.2.2 below, we examine the stability of these quasi-equilibria, and in section 4.3.2, the effect on them of changes in the exogenous variables.

4.2. Stability of the quasi-equilibria.

4.2.1. Stability in the simple formulation case.

The simple formulation of money price adjustments has been shown to lead to a quasi-equilibrium in one of the two regions, DC and NRC, depending on the relative magnitudes of the autonomous demand for output, $\bar{Y}$, and the full-employment output, $F(\bar{N}^S, \bar{R}^S)$.

(i) With $\bar{Y} < F(\bar{N}^S, \bar{R}^S)$, from Figure 4.4a, a quasi-equilibrium occurs in region DC of Keynesian unemployment and falling money prices at point Q_{SF}^{DC}. It is shown in Appendix C3 that with perfectly inelastic factor supplies the quasi-equilibrium will be stable. Furthermore, we show that if the production function is Cobb-Douglas (with diminishing returns to scale) the quasi-equilibrium Q_{SF}^{DC} is a stable node. (A node exhibits convergence without cycling; a focus exhibits convergence with cycling.)

(ii) With $\bar{Y} > F(\bar{N}^S, \bar{R}^S)$, from Figure 4.4b, a quasi-equilibrium occurs in region NRC of repressed inflation and rising money prices at point Q_{SF}^{NRC}. It is shown in Appendix C3 that with perfectly inelastic factor supplies, the quasi-equilibrium will be a stable node.

4.2.2. Stability in the alternative formulation case.

The alternative formulation of money price adjustments has been shown to lead to one of three possible quasi-equilibria, depending on the relative magnitudes of the exogenous variables.

(i) With $\bar{Y} < F(\bar{N}^S, \bar{R}^S)$, from Figure 4.6a, a quasi-equilibrium may occur in region SC of classical unemployment at point Q_{AF}^{SC}. It is shown in Appendix C4 that it is sufficient for stability that F_{NR} is zero. The necessary conditions for stability are not simple.

(ii) With $\bar{Y} < F(\bar{N}^S, \bar{R}^S)$, a quasi-equilibrium may occur in region DC of Keynesian unemployment and, from Table 4.4, falling money prices, at point Q_{AF}^{DC}. It is shown in Appendix C4 that this is a stable quasi-equilibrium.

(iii) With $\bar{Y} > F(\bar{N}^S, \bar{R}^S)$, from Figure 4.6b, a quasi-equilibrium will occur in region NRC of repressed inflation and, from Table 4.4, rising money prices, at point Q_{AF}^{NRC}. It is shown in Appendix C4 that this quasi-equilibrium is a stable node.

4.3. Comparative statics of the quasi-equilibria.

Section 2.3.2 analyzed the effects of exogenous disturbances on the general market clearing equilibrium of the basic model. To assure that clearing in all three markets was possible (that is, that the system of market-clearing conditions was consistent), we had to assume that the government altered its fiscal policy after the change in the exogenous variable(s), so that autonomous demand for output continued to equal full-employment output:

$$\bar{Y} = F(\bar{N}^S, \bar{R}^S).$$

In the absence of such action, no point of general market clearing would exist; that is, there would be no combination of real wage w and real resource flow price v such that supply equalled demand in all three markets simultaneously. In the absence of a general market clearing point, the basic model does not tell us much, since its underlying assumptions mean that it can only describe the adjustment of the tatonnement process: we assume in the basic model that no non-market-clearing trading occurs, or that any trading that does take place is recontractable after the general market clearing combination of w and v has been determined by the tatonnement. In the absence of a general market clearing combination of w and v, the virtual adjustment process of the basic model (as described in section 2.4.3) cannot result in trading. To overcome this obvious deficiency in the model's description of the real world, the non-market-clearing trading model of Chapter III was developed, and its price adjustments examined in this chapter.

Using the non-market-clearing model of Chapters III and IV, we no longer have to assume that the government's fiscal policy successfully achieves clearing on the output market. Trading will take place: labour will be hired, flow of resource bought, goods produced and sold, in the absence in general of market clearing on any of the three markets. Government fiscal policy becomes a true exogenous variable, with government expenditure G and taxation receipts T included in the definition of autonomous expenditure $\bar{Y}$, equation (2.27):

$$\bar{Y} \equiv (I + G - (1-s)T)/s.$$

We treat $\bar{Y}$ as an exogenous variable which can change. Other exogenous variables to be examined are resource supply $\bar{R}^S$ (to foreshadow the next chapter), and the production function, which can be subject to neutral or resource-augmenting technical change. We examine the comparative statics of the model under the two hypotheses for price adjustment: the simple formulation (SF), and the alternative formulation (AF).

4.3.1. Comparative statics with the simple formulation.

Under the simple formulation of price adjustment we have shown in section 4.1.1 above that there are two possible quasi-equilibria: iff $\bar{Y} < F(\bar{N}^S, \bar{R}^S)$ then Q_{SF}^{DC} occurs in the region of Keynesian unemployment with falling money prices, shown in Figure 4.4a; iff $\bar{Y} > F(\bar{N}^S, \bar{R}^S)$ then Q_{SF}^{NRC} occurs in the region of repressed inflation with rising money prices, shown in Figure 4.4b. In the case where $\bar{Y} = F(\bar{N}^S, \bar{R}^S)$, shown in Figure 4.3, there is a region, DNRC, where general market clearing occurs with all prices, real and nominal, constant. The calculations involved with the comparative statics of this section are shown in Appendix C5, and the results summarized in Table 4.5.

An increase in resource supply $\bar{R}^S$ is shown to lead to an unequivocal rise in the quasi-equilibrium real wage w* at Q_{SF}^{NRC}, and to a rise or fall in w* at Q_{SF}^{DC}, depending on the values of the exogenous variables. The real resource flow price v* will fall at Q_{SF}^{DC}, while it may rise (against intuition) or fall at Q_{SF}^{NRC}. Consider equation (C54),

$$(4.18) \qquad dv^* = -\lambda_N N_w^{RCD}(-\lambda_V + s\lambda_P F_R)\, d\bar{R}^S/|J| \gtrless 0.$$

From Table 3.3, N_w^{RCD} is negative, and from equations (C12) and (C22), $|J|$ is positive. Thus we see that if

$$(4.19) \qquad \lambda_V/\lambda_P < sF_R,$$

then dv* is positive at Q_{SF}^{NRC}. That is, if the ratio of the speeds of adjustment of the resource and output money prices is less than the fraction of the marginal product of resource saved, then an increase in the resource supply leads to an increase in the quasi-equilibrium v* at Q_{SF}^{NRC}. We shall show in section 4.3.2 that these results are similar for quasi-equilibria Q_{AF}^{DC} and Q_{AF}^{NRC}, respectively.

In economic terms, in region NRC an increase in $\bar{R}^S$ leads to increases in Y^{NRCS}, the effective supply of output (and the quantity of output traded), and in N^{RCD}, the effective demand for labour. These in turn lead to decreases in Y^X and R^X, and to an increase in N^X. With the simple formulation of price adjustment equations (4.7), this results in decreases in $\dot{P}/P$ and $\dot{V}/V$, and an increase in $\dot{W}/W$, which are all positive in NRC. Thus, for any value of v, the value of $\dot{w}/w$ increaes, and the

Table 4.5. The effects of exogenous change.

		$d\bar{R}^S > 0$	$d\bar{Y} > 0$	Technical progress	
				Neutral	Resource-augmenting
Basic model	dv*	–	na	+	– (large α)
	dw*	+	na	+	+
Q^{DC} (SF and AF)	dv*	–	±	±	±
	dw*	±	±	±	±
Q^{NRC} (SF and AF)	dv*	±	–	+	+ (large α)
	dw*	+	–	+	+
Q^{SC}_{AF}	dv*	–	±	±	±
	dw*	+	±	±	±

constant-w locus in Figure 4.4b moves up. For any value of w, the value of $\dot{v}/v$ will fall (and the constant-v locus move to the left) only if the decrease of $\dot{P}/P$ is less than the decrease of $\dot{V}/V$. The smaller λ_P and the larger λ_V, the more likely this is to happen. Inequality (4.19) gives a precise statement of this condition.

An increase in autonomous demand for output $\bar{Y}$ is shown to lead to unequivocal falls in v* and w* at Q_{SF}^{NRC}, and to movements of inconclusive sign of v* and w* at Q_{SF}^{DC}. These are results that hold also for Q_{AF}^{NRC} and Q_{AF}^{DC}, respectively.

The effect of neutral technical progress is to increase both v* and w* at Q_{SF}^{NRC} and to lead to movements of inconclusive sign for both v* and w* at Q_{SF}^{DC}. Resource-augmenting technical progress will lead to similar increases in v* and w* at Q_{AF}^{NRC} if the parameter α is large; otherwise v* might fall: low $\alpha > 1$ is more likely at the outset of technical change. In the case of Q_{SF}^{DC}, resource-augmenting technical change can result in a fall or a rise of the quasi-equilibrium real prices v* and w*. These results hold also for Q_{AF}^{NRC} and Q_{AF}^{DC}.

Table 4.5 summarizes these results, and shows that dropping the unrealistic assumption of the basic model of no non-market-clearing trading has resulted in some changes. Except for dv* in response to an increase in $\bar{R}^S$, the comparative statics of the basic model are seen to be identical to those of Q_{SF}^{NRC} and Q_{AF}^{NRC}. But the comparative statics of Q^{DC} and Q_{AF}^{SC} are different. In particular note that an increase in government expenditure, which will result in an increase in autonomous demand for output $\bar{Y}$, will lead to a <u>fall</u> in v* and w* at Q^{NRC}, and may lead to a fall at Q^{DC} and Q_{AF}^{SC}.

In region NRC an increase in $\bar{Y}$ leads only to an increase in Y^X. With the simple formulation of price adjustment equations (4.7), this results in an increase in $\dot{P}/P$. Since $\dot{V}/V$ and $\dot{W}/W$ are unaffected, the constant-v locus shifts to the left in Figure 4.4b, and the constant-w locus shifts down: both w* and v* fall at Q_{SF}^{NRC}. In region DC an increase in $\bar{Y}$ leads to increases in N^{DCD}, the effective demand for labour (and level of employment), and in R^{DCD}, the effective demand for resource flow (and level of resource use). These in turn lead to increases in Y^X, N^X, and R^X, which are all negative. With the simple formulation of price adjustment, this results in increases in $\dot{P}/P$, $\dot{W}/W$, and $\dot{V}/V$, which are all negative in region DC. For any value of w the value of $\dot{v}/v$ will rise (and the constant-v locus move to the right) only if the increase of $\dot{P}/P$ is less than the increase in $\dot{V}/V$. Similarly for $\dot{w}/w$ and the constant-w locus. Thus, depending on the relative magnitudes of the changes in $\dot{P}/P$, $\dot{W}/W$, and $\dot{V}/V$, v* and w* can rise or fall at Q_{SF}^{DC}. This is stated in equation (C48).

In the case of $\bar{Y} = F(\bar{N}^S, \bar{R}^S)$ shown in Figure 4.3, a change in one or more of the exogenous variables will lead in general to

$$\bar{Y} \neq F(\bar{N}^S, \bar{R}^S).$$

Thus an increase in $\bar{R}^S$ or technical change of either kind will lead to

$$\bar{Y} < F(\bar{N}^S, \bar{R}^S)$$

and the situation shown in Figure 4.4a. An increase in $\bar{Y}$ will lead to

$$\bar{Y} > F(\bar{N}^S, \bar{R}^S)$$

and the situation shown in Figure 4.4b. In the case of the alternative formulation of price adjustment, the situation shown in Figure 4.5 will become that of Figure 4.6a or 4.6b, respectively.

4.3.2. Comparative statics with the alternative formulation.

As noted in the previous section, the comparative statics of Q_{AF}^{DC} and Q_{AF}^{NRC} are identical with those of Q_{SF}^{DC} and Q_{SF}^{NRC}, respectively. They are summarized in Table 4.5. We can say little about the movement of Q_{AF}^{SC} in response to changes in exogenous variables: for technical change of either kind and for changes in autonomous demand for output, both v* and w* can rise or fall depending on the values of the exogenous variables and parameters of the model. But in the case of an increase in $\bar{R}^S$, we show in Appendix C6 that v* will fall and w* will rise at Q_{AF}^{SC}, the quasi-equilibrium of classical unemployment, shown in Figure 4.6a. Symmetrical analysis would show that an increase in labour supply $\bar{N}^S$ would result in an increase in v* and a fall in w*, as the classical analysis of the general market clearing basic model would predict.

4.4. Summary of Chapters III and IV.

We have developed a model which does not include the assumptions of recontracting or of no non-market-clearing trading. This disequilibrium, short-run model includes two previously symmetrical variable factor inputs, labour and resource flow. We have introduced the concept of quasi-equilibrium of Hansen (1951), and examined the existence, uniqueness, and stability of possible quasi-equilibria under two formulations of price adjustment, both ad hoc variations of the Walrasian excess (effective) demand hypothesis. Quasi-equilibria under the simple formulation were found to be matched by similar quasi-equilibria under the alternative formulation in the DC and NRC regions of Keynesian unemployment and repressed inflation, respectively, although this formulation also allowed a quasi-equilibrium in the region SC of classical unemployment, which did not occur under the simple formulation. The comparative statics of the quasi-equilibrium were examined. We shall refer to these results when considering the longer run in Chapter VI, since they indicate how the economy would behave in the longer run.

In the following chapter we shall examine the behaviour of the resource suppliers, who are assumed to hold resource stock as one asset in a portfolio. We shall examine the implications of their expectations and behaviour on the stability of the model, and in particular we shall look for micro-foundations for the Hotelling principle of efficient, intertemporal allocation of non-renewable resources. We shall examine the possibility of harmony between the Hotelling principle's holding in the resource flow market and the quasi-equilibrium in the labour market--this will conclude the answer to the question of whether our model can support equilibrium in both the stock (asset) and flow markets for resource, and if so, how.

We shall not continue using the alternative formulation of price adjustment. We have shown above that it results in behaviour of the model similar to that with the simple formulation. We shall examine specifically the two quasi-equilibria found in this chapter to occur with the simple formulation: Q_{SF}^{DC} and Q_{SF}^{NRC}. It is possible that a formulation which explicitly included price inflation expectations, such as the Phelps-Friedman formulation, would result in different, more realistic, behaviour of the model, but examination of this will have to await further work.

CHAPTER V: EXPECTATIONS AND THE SUPPLY OF RESOURCE FLOW

5.1. The Hotelling principle.

So far, in Chapters II, III, and IV, the two variable factor inputs, labour services and resource flow, have been treated symmetrically, although in section 2.2.3 we introduced the idea that the flow of resource, supplied for sale to firms by the owners of the stocks of non-renewable natural resource, was a decreasing function of their expectation of future prices of the resource stocks. The existence of stocks of resource, which is not merely used in production (like capital, for which wear-and-tear is a minor characteristic), but which is used up in production and cannot be reproduced (a distinction echoing Kaldor's (1939/40) terms, "Gebrauchsgüter" and "Verbrauchsgüter", or "fixed-capital-goods" and "working-capital-goods", respectively) means that the two factor inputs are fundamentally different, for although in the short run the total labour force can be thought of as fixed, we do not think of labour as being used up, and it is, of course, reproducible.

Stocks of non-renewable natural resource (which is exhausted in production) are assets to society and their owners, much like reproducible assets. Although (apart from their value as objects of speculation, or their believed intrinsic value to society, if these exist) their ultimate value lies in their potential for being transformed in the production process with other factor inputs into scarce output, they can be bought and sold as stocks, and can appreciate or depreciate, as can other assets. In fact, certain resources may be traded on two distinct markets, a stock market, in which the object of buying is to realise future anticipated capital gains, and a flow market, in which the object of buying is to obtain quantities of resource as a factor input to production. We have implicitly drawn this distinction in the use of the phrase "resource flow." Two distinct markets implies two distinct prices, but we shall assume that arbitrage between the markets (if they exist separately) leads to equality of the prices. We shall continue to denote the nominal (money) price of resource as V, and the real price of resource as v.

If owners of resource stocks regard them as perfect substitutes for other, dividend-earning, assets, then we can speak of a single asset market, on which resource owners can trade stocks of resource and other assets. As Solow (1974a) notes, asset markets can only be in equilibrium when all assets in a given risk class earn the same rate of return, partly as current dividend, and partly as capital gain. We assume that all other assets in the economy can be modelled as one, with the common rate of return of the interest rate, denoted as r. The only way, apart from convenience for use, that a stock of resource can produce a current return for its owner is by appreciating in value. Then equilibrium in the asset market will only occur when the value of a resource stock is growing at a rate equal to the interest rate r. Since the value of a stock is also the present value of future sales from it (after deduction of extraction costs), in asset market equilibrium resource owners must expect the net price of the resource to be growing at a proportional rate equal to the rate of interest. (In a competitive market net price equals market price minus marginal cost; under a monopoly net price equals marginal profit, which equals marginal revenue minus marginal cost.) We consider a competitive asset/resource stock market, and we abstract from explicit extraction costs. Thus, in continuous time formulation, the necessary condition for asset market equilibrium can be written as

$$(5.1) \qquad e = r,$$

where e is defined as being the expected proportional rate of change of real resource price at any time, from equation (2.17),

$$(5.2) \qquad e(t) \equiv \dot{v}(t)^{e}/v(t).$$

This fundamental principle of the economics of exhaustible resources is known as the Hotelling principle after the author of the first paper to present it (Hotelling (1931)). He thought of it as a condition, not of stock equilibrium on the asset market, but of flow equilibrium on the resource market: if the price of resource v is growing at a proportional rate equal to the interest rate, then owners of stocks of resource will be indifferent at the margin between (extracting and) selling, and holding stocks at every instant of time.

Several authors have followed Hotelling in considering the behaviour of owners of resource stocks acting to maximize the present value of their stocks (Gordon (1967), Herfindahl (1967), Cummings (1969)). They have in general agreed that necessary conditions for this maximization include the Hotelling principle, where the net price (or marginal user cost) grows at the rate r so that holding unused-up resource is as attractive as other investments, and a marginal cost condition governing the rate of extraction, which states that the price equals the marginal extraction cost plus the marginal user cost (or net price, or marginal profit, or opportunity cost of using the limited resource now rather than later). Given the general case of non-zero marginal extraction costs, Peterson and Fisher (1977) note that the price will follow a more complicated path than exponential growth, depending on the rates of change of marginal extraction and user cost, which in turn depend on both supply and demand factors: price could easily fall as a result of technical progress, the discovery of new deposits, or (significantly) a slackening of demand.

Assuming that the interest rate is equal to the rate of social time discounting, several authors have examined the question of optimal resource depletion in the context of optimal growth theory, with a hypothetical planner attempting to maximize some more general social welfare function (Koopmans (1973), Garg (1974), Garg and Sweeney (1978), Dasgupta and Heal (1974), Solow (1974b), Stiglitz (1974a)). They determine the conditions under which optimal programs and steady states exist, and derive properties of the optimal paths. In several cases the Hotelling principle is demonstrated to be a necessary (although not sufficient) condition for the maximization of discounted future utility from consumption. Other writers have introduced uncertainty in the size of the stock (Gilbert (1976), Loury (1976)), the existence of a "backstop" technology as a substitute for the resource (Nordhaus (1973), Dasgupta and Stiglitz (1975)), and increasing costs of extraction (Heal (1976), Levhari and Liviatan (1977)). The Hotelling principle continues to play an important role in each of these cases.

After the demonstration that, even in an economy in which natural resources are essential and exhaustible, there may, under certain conditions, exist sustained growth in *per capita* income (Stiglitz (1974a)), the question arises of whether this socially optimal allocation can be supported as a competitive market allocation. What will the properties of the time paths generated by a market of present-value-maximizing resource owners and consumers be? In the absence of the usual market imperfections, it is well known that a competitive equilibrium of present value maximizers is Pareto-optimal, but can this result be extended to an economy including non-renewable natural resources? Sweeney (1977) shows that, in the absence of the usual market imperfections, if the savings/consumption time paths are chosen optimally, the socially optimal allocation can be supported as a competitive allocation, under appropriate convexity conditions, assuming perfect foresight and no uncertainty. In the case of zero extraction costs, the Hotelling principle implies that extraction rates are chosen so that the price rises at the interest rate.

Weinstein and Zeckhauser (1975) closely examine market equilibrium in a competitive model. They demonstrate that if resource producers (or owners) have perfect foresight of future demands for resource flow or if resource consumers (the firms in our models) can store the resource at zero cost, equilibrium over time can only occur with time paths obeying the Hotelling principle: with zero extraction costs, no resource flow will be sold in period t when

$$(5.3) \qquad v_t < \max_{\tau}(v_{\tau}(1+r)^{t-\tau}),$$

that is, when the real price of resource in period t, v_t, falls below the highest trajectory, and none will be bought in any period t for which there exists an earlier period $\tau < t$ such that

(5.4) $$v_t > v_\tau (1+r)^{t-\tau},$$

that is, when the real price of resource in period t rises above the lowest trajectory. Solow (1974a) characterizes this as demand for resource flow just equal to its supply at the current price, with clearing in the market for resource flow: "No other time profile for prices can elicit positive production [of resource flow] in every period of time" (1974a, p. 3). If the net price rose too slowly, supply of resource flow would be made earlier, and the resource would be exhausted more quickly, since no one would want to hold resource stocks and earn less (in capital gains) than the going rate of return on other assets. If the net price rose too fast, holding resource stocks would be an excellent investment, and owners would postpone selling while they enjoyed super-normal capital gains.

In the case where resource producers (or owners) do not have perfect foresight of future demands for resource flow, and where resource consumers (or firms) face infinite storage costs, and in the absence of a futures market, future demand for resource flow is uncertain. Weinstein and Zeckhauser (1975) show that for risk-neutral resource suppliers the market equilibrium condition is that the expected resource price grows at a rate equal to the interest rate, the Hotelling principle as stated above in equation (5.1). Both Nordhaus (1973) and Stiglitz (1974b) note that the condition of equilibrium in the asset market (the Hotelling principle) results in a family of price trajectories (with zero extraction cost), each having different levels of price at any instant. The unique solution of resource depletion depends on the terminal solution that all resources are used up at the end of (and not before) the last period. If the time horizon is at infinity, this becomes the transversality condition of optimal control theory.

Stiglitz (1974b) has shown that even if resource producers possess myopic perfect foresight:

$$(5.5) \qquad e = g,$$

with the expected rate of change of resource price at any time equal to the actual rate of change of resource price (and equal to the rate of interest if the Hotelling principle holds), without either infinite perfect foresight or a complete set of futures markets extending infinitely far into the future, there exists no economic mechanism which will guarantee that the initial price is set so that the transversality condition is satisfied and the economy converges to balanced growth: if the initial price of the resource is set too low, elastic demand will result in the resource stock's being used up in finite time; if the initial price is too high, there will always be a finite amount of resource stock remaining--an inefficient situation with over-saving of resources. This long-run instability of the economy in the absence of future markets is of the same kind that Hahn (1966) and Shell and Stiglitz (1967) have shown can exist in growth models with more than one capital good, which is essentially the same as has often been attributed to speculative markets, where the future plays an important part in determining current decisions, but forward markets do not.

But we are concerned with the short-run stability of an economy including non-renewable resources. In particular we examine the microeconomic foundations of the economy's moving along a long-run price trajectory, with the asset market in equilibrium and the Hotelling principle's being obeyed. We have noted above that the Hotelling principle has been derived as a necessary condition for socially optimal intertemporal allocation of a non-renewable resource. We have also noted that it has been argued that, with zero extraction costs, resource owners acting to maximize the present value of their stocks must extract at such a rate that the Hotelling principle holds. Finally, several authors (Stiglitz (1974b), Weinstein and Zeckhauser (1975), and Sweeney (1977), inter alia) have argued that in a competitive economy of present value maximizers, with zero extraction costs, the Hotelling principle will be obeyed by the economy. These general equilibrium models are so complex that in order to derive intelligible results it is necessary to make very strong assumptions, as has been done, in particular that arbitrage between resource stock and other asset markets occurs so rapidly that the rates of return on all assets are always equal, and that markets clear instantaneously.

But if the resource market clears at every instant, what is the driving force behind the rise in price? In Chapter IV we examined several theories of price adjustment, and the consequences of each for the fix-price disequilibrium model of Chapter III. In this chapter we introduce adjusting expectations of future resource price and examine the existence, uniqueness, and stability of short-run equilibria and quasi-equilibria in the resulting model. In particular we shall ask how equilibrium in the asset market (resulting in the Hotelling principle) could be reconciled with short-run equilibrium or quasi-equilibrium in the system of three flow markets examined in Chapters III and IV. We shall at first assume no autonomous price change: the money price on a market will change only in response to an imbalance of supply and demand on that market; that is, f(0) = 0 in the Walrasian excess demand equation (4.1). We shall later examine the consequences for the model of allowing expectation of price change to affect actual price change directly.

The model developed in the previous chapter allows us to relax the strong assumptions made in other studies of general equilibrium models: that arbitrage between resource stock and other asset markets occurs so rapidly that the rates of return on all assets are always equal, and that markets clear instantaneously. We shall find that despite the present-value-maximizing behaviour assumed of resource suppliers, there is no basis for belief that such economies will obey the Hotelling principle, under reasonable assumptions.

One other study has attempted to look at these questions. In a paper which examines the consequences for exhaustible resource allocation of the non-existence of forward markets and contingent commodity or risk markets, Heal (1975) builds three models to examine the resource depletion rates that will occur under plausible adjustment mechanisms for price and quantity in disequilibrium situations. The simplest model, in our symbols, is

$$(5.6) \qquad \dot{v} = R^D - R^S$$

$$U'(R^D) = v,$$

$$\dot{R}^S = v - \bar{v}^e.$$

The first equation states that the current resource price adjusts at a rate proportional to the (notional) excess demand for resource (flow). The second states that marginal utility (or payoff) of resource consumption equals the price, where the consumption equals the demand for resource. The third states that the (flow) supply of resource changes at a rate proportional to the difference between the current price and the expected future price, which is static or exogenous: an increase in the current price above the expected calls forth an increase in supply as sellers take advantage of attractive market conditions, and vice versa. If the utility function is logarithmic the system (5.6) can be written

$$(5.7) \qquad \dot{v} = 1/v - R^S$$
$$\dot{R}^S = v - \bar{v}^e,$$

which converges to a solution where $R^S = 1/v$ and $v = \bar{v}^e$ (demand equals supply and the price is at its (exogenous) expected level). Heal himself argues that in any market expected future prices will probably be influenced by the history of past prices, and, in the long run, the balance between future supply and demands of resource will be influential. In addition, we can see that the model ignores the problem of voluntary exchange out of equilibrium: if $R^S < R^D$ then consumption will be R^S, not R^D.

Heal's second model incorporates adaptive expectations:

$$(5.8) \qquad \dot{v}^e = \alpha_1(v - v^e) + \alpha_2 R^S,$$
$$\dot{v} = 1/v - R^S,$$
$$\dot{R}^S = v - v^e.$$

The first equation states that expectations are revised upwards if the present price exceeds that which had been expected for the present, and vice versa, and that a decline in the reserve-production ratio boosts price expectations. The only equilibrium to this system occurs when $R^S = 0$, $v = v^e = \infty$, that is, when the resource stock has been exhausted. Heal criticises the models (5.6) and (5.8) for their partial equilibrium nature. As we have outlined in previous chapters, interactions between resource markets and the remainder of the economy are important, both because traders arbitrate between resource stocks and other assets, and because the price of resource influences the activity elsewhere in the economy, which in turn may affect the demand for resources.

Heal's third model is a compromise between the general and partial equilibrium approaches:

$$\text{(5.9)} \qquad \begin{aligned} \dot{e}/e &= \beta(g-e) + \gamma\dot{R}^S/R^S \\ \dot{v}/v \equiv g &= -(1/\eta)\,\dot{R}^S/R^S, \\ \dot{R}^S/R^S &= \alpha(r-e), \end{aligned}$$

where η is the elasticity of demand for resources and r is the rate of return available from holding assets other than resources. The first equation of (5.9) is similar to the adaptive expectations model of (5.8), except that it is stated in terms of rates of change rather than levels: this is required by the fact that it is the rate of change of the resource price that determines the return (capital gain) from holding it. The second equation follows from the assumption that the resource market clears instantaneously, with $R^S = R^D(v)$. The third equation embodies the idea that there is arbitrage between resource and capital markets, but that this is not sufficiently rapid to produce complete equatization of returns: if the rate of return r elsewhere exceeds the expected rate of capital appreciation on resource e then resource owners sell resources to hold claims on real capital, to an extent depending on the difference between r and e. It can be shown that for "low" elasticity of demand this system exhibits unstable, saddle-point, behaviour. It has an equilibrium trajectory with constant proportional growth in the expected price (that is, constant e) greater than the interest rate r, and actual prices growing at a constant rate, with the level of consumption falling exponentially so that the reserve production ratio remains constant. This trajectory mimics the long-run behaviour of an optimal depletion path, and is an unstable path. If the elasticity of demand is "too great," then the path is stable with the expected price growing at a constant rate less than r, prices falling, and consumption rising exponentially.

A strong criticism of all of Heal's models is the fact that the behaviour of resource price is strongly dependent on the characteristics of the resource demand function and the fact that the resource market is assumed to clear instantaneously. Heal has not been able to escape sufficiently from the optimal growth models of the earlier studies to develop short-run models. In contrast the model presented in Chapters III and IV does not assume a specific demand function, nor does it assume that the resource flow market clears. It is a stronger contender for Heal's aim to build a general equilibrium model of the stability of an economy with exhaustible resources, including as it does three explicit markets (output, labour, and resource flow) and the possibility of non-market-clearing trading. It remains to include explicitly the present-value-maximizing behaviour of the resource owners as they form expectations of the future prices of resource and respond accordingly.

5.2. Expectations and resource supply.

We want to define a single, reasonable, variable to describe expectations of prices. Our definition should reflect that this variable will be a link between the supply of resource and the past and present behaviour of the system. Resource supply is a function of expectations since no futures markets are assumed to exist for the natural resource. (This is a polar case, but will allow attention to be focused on the consequences of different modes of expectation formation, expectations which having been formed without forward markets might then be thought of as being revealed by such markets if they existed.)

It is assumed that suppliers of resource, in deciding whether to hold stocks or to sell them, are comparing the return they expect to receive from holding them with the return they could receive from holding other assets. The only possible return from holding stocks of natural resource is a capital gain, stemming from price increases, but of course in making their decision of whether or not to sell at the margin, resource owners are responding to expected capital gains: not until the actual sale takes place is the capital gain (or loss) realised and their expectations shown to be correct, or optimistic, or pessimistic.

Hence the adoption of the proportional rate of change of the expected price as the "expectation" variable by many writers on natural resource economics. In discrete formulation this is written as

$$(v^e(t+1) - v(t))/v(t) \tag{5.10}$$

where $v^e(t+1)$ is the expected price at time $(t+1)$, and where $v(t)$ is the (known) price at time (t). In continuous time we write e for the variable, where, from equation (2.17),

$$e \equiv \dot{v}^e/v \tag{5.11}$$

where $\dot{v}^e$ is the time rate of change of expected real price. Real price is used here because unless suffering from money illusion the resource owners are interested in the real, not nominal, expected capital gain; in this case the return they could receive from holding other assets must be expressed in real terms; that is, if the return from holding other assets is expressed as an interest rate, it must be deflated for the rate of expected inflation of money prices, to result in Fisher's real rate of interest.

For simplicity we assume that all individuals form their expectations in the same way, and that the expected rate of change of price is held uniformly by all participants in the asset market. There is no compelling reason to believe that this is true--experience would suggest that it is not--but the effect of non-uniform expectations can be modelled with a continuous supply of resource flow function: if e is higher, ceteris paribus, supply is lower, but not discontinuously since e can be thought of as a proxy for the distribution of expected price change across the asset market participants, and some, who have lower expectations than the median , will still sell.

If each resource supplier treats the prices of natural resources parametrically, both the spot price and expected prices, that is, if each resource owner is a price-taker, then in a Hotelling world if the proportional rate of expected price, e, is greater (less) than the real interest rate, r, then no resource owner will want to sell (hold) any resource stocks, since the expected return from holding resource stocks is greater (less) than the expected return from holding other assets. This behaviour leads to a horizontal resource supply curve at any instant. Of course, this behaviour on the part of the resource owners is a function of the definition of the "expectation" variable and the mode of expectation formation: if an increase in the spot price at any moment above the previously expected level leads to an increase in e, will this be seen as a sign that e will increase still further, thus encouraging holding, or will it be seen as a temporary variation from a lower "normal" level, thus encouraging selling? (Intuition suggests that either expectation will tend to be self-fulfilling.) The relationship between e and the spot price and past prices will have to be formulated so as to model these two possibilities.

5.2.1. Two resource supply functions.

We shall examine two possibilities for $R^S(e)$, the supply of resource flow as a function of the proportional rate of change of expected price:

(i) a simple, downwards-sloping function as an approximation to the Hotelling world; as the expected rate of change of price increases, ceteris paribus, the supply of resource flow drops. This continuous function, seen in Figure 5.1, can be thought of as modelling the effects of non-uniform expectations across the asset market participants: if e is thought of as the average expected rate of change of price, in some sense, then there will always be some individuals on the lower tail of the distribution of expected price change who will sell, for high e, and likewise some on the higher tail who will not sell, for low e. Hence is obtained a smooth, downwards-sloping supply function. We shall denote it as

$$(5.12) \qquad R^S = R^{SS}(e), \; dR^{SS}/de < 0.$$

(ii) A function which more closely models the sell-all/hold-all Hotelling world described above; the suppliers' behaviour becomes limited with a "floor" minimum flow with high expected price change and with a "ceiling" maximum flow with low expected price change, as seen in Figure 5.2. Formally, the supply function can be written

$$(5.13) \qquad R^S = R^{HS}(e) \equiv \begin{cases} R_2^S \quad , & e \leq r - a, \quad a > 0 \\ R_1^S < R_2^S, & e \geq r + a \end{cases}$$

$$dR^{HS}/de = -b < 0, \quad -a < e - r < a.$$

This is not strictly a true Hotelling-type function, since, in a Hotelling world with uniform expectation of future price change, when e > r (the expectation of the change of real resource price is greater than the interest rate on alternative assets), stocks of resource deposits are seen as an excellent way of holding wealth, and owners of these stocks delay sale while they anticipate enjoying supernormal capital gains; and when e < r, stocks of resource are seen as an uncompetitive way of holding wealth, with owners of these stocks selling as much as they can in order

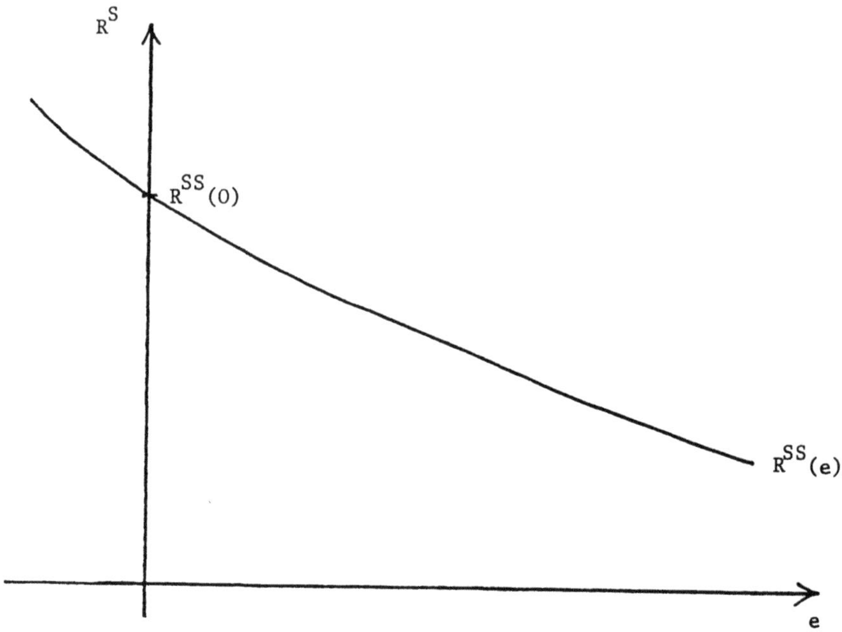

Figure 5.1: A simple resource flow supply function, $R^{SS}(e)$.

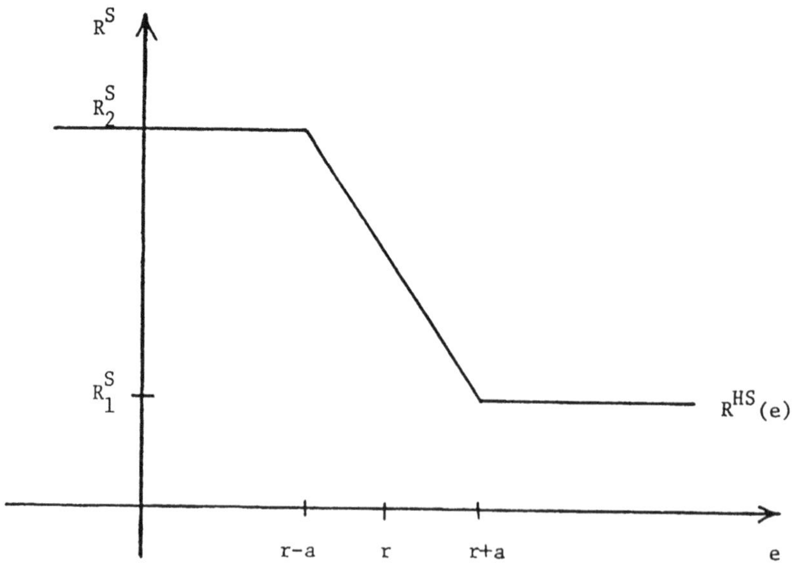

Figure 5.2: A quasi-Hotelling resource flow supply function, $R^{HS}(e)$.

to invest in assets yielding higher returns. At the limit no resources would be for sale when $e > r$, and all stock would be for sale when $e < r$. The formulation above tends to the limit as $a \to 0$, $R_2^S \to \infty$, and $R_1^S \to 0$. It allows us to examine the limiting case, but also the case of finite a, R_2^S, and R_1^S. We can think of the finite upper limit R_2^S on resource flow as implicitly embodying increasing costs of extraction: rather than being a physical limit it is an economic limit. In the short run it is unprofitable for the resource owners to supply more rapidly than R_2^S because of increasing costs of extraction. The lower limit R_1^S and the range $(-a < e - r < a)$ in which the resource flow supply is elastic can be thought of as resulting from incomplete knowledge of investment opportunities or from non-uniform price expectations among resource suppliers.

5.2.2. Five modes of expectation formation.

We shall examine five possibilities for the mode of expectation formation:

(i) Static expectations, in which, despite variations in the actual rate of change of real resource price, resource suppliers continue to expect an exogenous, uniform, constant rate of change $\bar{e}$. This situation, one of no learning, was introduced in section 2.2.3, equation (2.18),

$$e = \bar{e}, \tag{5.14}$$

and has been thoroughly examined in Chapters II, III, and IV.

(ii) Myopic perfect foresight, in which the proportional rate of change of expected price equals the actual rate of change of price, the case in which expectations are realised at every instant of time, which can be written

$$(5.15) \qquad e = g, \qquad \text{where}$$

$$(5.16) \qquad e \equiv \dot{v}^e/v, \text{ the expected rate of change of price, and}$$

$$g \equiv \dot{v}/v, \text{ the actual rate of change of price.}$$

This is a special case of the next formulation.

(iii) The classical model of simple adaptive expectation formation, where individuals revise upwards their expectations of the rate of change of prices if the actual rate of change exceeds what had been expected, and vice versa. This can be written

$$(5.17) \qquad \dot{e} = \mu \cdot (g - e), \qquad \mu > 0.$$

This is related to the next formulation.

Clearly, how people form their expectations of future prices is a complex matter, and one in which individual expectations can be expected to differ. For simplicity we have assumed above, and shall continue to do so, that the expected rate of change of price is held uniformly by all participants in the asset market, although non-uniform expectations can be modelled in the resource supply function. Essentially, the problem is how, in the absence of futures markets, to calculate market-clearing prices for future dates. Thus it makes sense to model expectation formation not only on past prices, but on current and past values of real magnitudes, although we shall confine our formulation to prices, since in the short run real magnitudes do not change greatly. But even focusing only on prices in the adaptive expectations model cannot capture the phenomenon known as "regressive expectations": if prices rise slightly from their "normal" level, adaptive expectations extrapolates this price increase, but it is equally plausible that some individuals expect the price to return to its previous level. The problem is the formulation of some simple but "reasonable," analytically tractable hypotheses of expectation formation, and to investigate their implications for the stability of the economy.

Stiglitz (1974b) was able to find only three basic properties which "reasonable" expectation models ought to have. If, in discrete time, we write

(5.18) $$v^e(t+1) = \phi(v(t), v(t-1), \ldots, v(t-n), \ldots; t)$$

then he argues that ϕ should have the properties of

(a) linear homogeneity:

(5.19) $$\lambda v^e(t+1) = \phi(\lambda v(t), \lambda v(t-1), \ldots; t)$$

(b) stationarity, i.e., ϕ is independent of t

(c) if prices have never changed, then the expected price is the present price

(5.20) $$\phi(v, v\ldots) = v.$$

Stiglitz also makes the useful postulate that the individual summarizes past information into a weighted average of past prices, or "normal" price:

$$(5.21) \qquad \bar{v}(t) \equiv \beta \int_{-\infty}^{t} v(x).\exp^{-\beta(t-x)} dx, \qquad \beta > 0,$$

which by differentiation leads to

$$(5.22) \qquad \dot{\bar{v}} = \beta.(v - \bar{v}).$$

(iv) Hence the formulation that expected future prices are a function of present and average past prices

$$(5.23) \qquad \dot{v}^{e} = \hat{\psi}(v, \bar{v}).$$

Linear homogeneity, assumption (a), implies that we can write

$$(5.24) \qquad e \equiv \dot{v}^{e}/v = \psi(\bar{v}/v).$$

Assumption (c) implies that

$$(5.25) \qquad \psi(1) = 0.$$

We can see that this formulation is consistent with both progressive and regressive expectations, depending on the sign of ψ': $\psi' > 0$ implies that if the present price is above the long-run average, the price is expected to fall, and vice versa (regressive expectations); $\psi' < 0$ implies that if the price is above "normal," it is expected to increase, and vice versa (progressive expectations).

Another version of the price level expectation formulation is

$$\dot{v}^e = \alpha.(v - \bar{v}). \tag{5.26}$$

Then $\alpha < 0$ corresponds to regressive expectations, and $\alpha > 0$ corresponds to progressive expectations. Division by v gives

$$e = \alpha.(1 - \bar{v}/v) \tag{5.27}$$

and differentiation and manipulation give

$$\dot{e} = \alpha g - ge - \beta e \tag{5.28}$$

or

$$\dot{e} = \alpha.(g - \frac{\beta+g}{\alpha}.e) \tag{5.29}$$

which is similar to the simple adaptive formulation, equation (5.17), above.

But although this formulation takes account of price levels in the immediate past, it does not take explicit account of the rate of change of prices in the immediate past. In a static market, such as the flow market has been modelled, where equilibrium occurs with constant (real) prices, this is not a shortcoming, but in a market where equilibrium occurs with constant proportional rate of change of price, the history of the rate of change of price in the immediate past may be expected to play a part in the formation of expectations. Note that equilibrium in the market for stocks of resource is assumed to occur when g = r, when the proportional rate of change of real resource price equals a constant, the interest rate or return from holding other assets, including for example pure consumer loans.

(v) Hence we postulate the formulation that expected future prices are a function of current rate of change, the equilibrium rate of change, and the expected rate of change of current prices. A particular formulation of this is the compound adaptive expectation formulation

$$\dot{e} = \gamma.(r - g) + \mu(g - e), \qquad \mu > 0, \tag{5.30}$$

which states that, if expected rate of price change equals actual at any moment, $\gamma > 0$ corresponds to regressive expectations in the sense that if the actual rate of change is less (more) than the interest rate, the expected rate of change increases (decreases), modelling the expectation that the actual and the expected rates of change will tend toward the stock market equilibrium of interest rate equals actual equals expected rate of change of price; $\gamma < 0$ corresponds to the progressive expectation that the stock market equilibrium is a knife-edge, and that any deviation from it will be amplified. Writers on the topic have disagreed on whether in fact the stock market equilibrium is stable: Stiglitz (1974b) emphasized its instability, and Kay and Mirrlees (1975) its stability. With explicit formulation of progressive or regressive expectations, we can examine whether the stability, or not, is a function of the expectations of the traders in the market.

To choose between the expectation of price level formulation, equations (5.26) and (5.28), and the compound adaptive expectation formulation, equation (5.30), we must ask which is more realistic in modelling the formation of expected future prices by the resource suppliers. To the reader who has only just been introduced to the exponential price rise of the Hotelling principle, the compound adaptive expectation formulation with its tendency for expectations to appraoch the interest rate may seem rather strange: in normal markets there is equilibrium with constant price level, and the price level formulation may seem a better model. But, as we have argued in section 5.1 above, rational resource suppliers will be concerned with the current return to and opportunity cost of holding stocks of non-renewable resource. The compound adaptive expectations formulation attempts to model this behaviour, a case perhaps of economic theory (that of exhaustible resources) affecting the behaviour of economic actors, who forecast using Hotelling-type models. The price level formulation models the behaviour of ignorant (or deluded) resource suppliers, who are not aware of the capital gains possible from holding stocks of resource. We assume below that modes of expectation formation are held uniformly by resource suppliers. We can examine the consequences for the economy of suppliers' having different models of expectation formation.

5.3. Stability of the quasi-equilibria.

The rest of the chapter is involved with examining the consequences of each combination of the two resource supply functions and the four modes of expectation formation introduced in section 5.2 on the stability of the three quasi-equilibria possible with the simple formulation of price adjustment of section 4.1.1. If, under reasonable hypotheses of the behaviour of the present-value maximizing resource suppliers, the quasi-equilibrium remains locally stable, we can conclude that we have shown that the micro-economics of our general disequilibrium model will not support the Hotelling principle in the short run: if the quasi-equilibrium is locally stable the real prices, and in particular the real price of resource v, will tend to the constant levels (w*, v*) characterizing the quasi-equilibrium. An economy obeying the Hotelling principle with the resource price v growing exponentially at the interest rate would be unstable, and would approach the appropriate quasi-equilibrium.

The simple formulation of price adjustment assumes that the money price in each of the three markets is a simple increasing function of the excess effective demand on that market. It is given by equation (4.7)

$$(5.31) \qquad g \equiv \dot{v}/v = \lambda_V R^X - \lambda_P Y^X$$

$$h \equiv \dot{w}/w = \lambda_W N^X - \lambda_P Y^X,$$

where it must be remembered that

$$(5.32) \qquad R^X = R^D - R^S(e)$$

since expectations are no longer static, but provide a link between the present and the unknown future, in the absence of forward markets. If we characterise the level of expected resource price rise at quasi-equilibrium as e*, then $R^S(e^*)$ is the supply of resource flow at quasi-equilibrium, and we can distinguish three possible quasi-equilibria.

(i) As shown in Figure 4.4a, iff

$$(5.33) \qquad \bar{Y} < F(\bar{N}^S, R^S(e^*)),$$

then Q^{DC} occurs in the region of Keynesian unemployment with falling money prices.

(ii) As shown in Figure 4.4b, iff

$$(5.34) \qquad \bar{Y} > F(\bar{N}^S, R^S(e^*)),$$

then Q^{NRC} occurs in the region of repressed inflation with rising money prices.

(iii) As shown in Figure 4.3, iff

$$(5.35) \qquad \bar{Y} = F(\bar{N}^S, R^S(e^*)),$$

then there is equilibrium with no change in money prices at the triple-market-clearing point DNR, Q^{DNR}.

The two possible supply of resource flow functions introduced in section 5.2.1 above are:

(i) The simple, downwards-sloping resource flow supply function shown in Figure 5.1 and formalized in equation (5.12)

$$(5.36) \qquad R^S = R^{SS}(e), \quad dR^{SS}/de < 0.$$

(ii) The quasi-Hotelling resource flow supply function shown in Figure 5.2, embodying implicit extraction costs and incomplete knowledge or non-uniform price-expectations, was formalized in equations (5.13)

$$(5.37) \qquad R^S = R^{HS}(e) \equiv \begin{cases} R_2^S & , \; e \leq r - a, \quad a > 0 \\ R_1^S < R_2^S, & e \geq r + a \end{cases}$$

$$dR^{HS}/de = -b < 0, \quad -a < e - r < a.$$

There are five possible modes of resource price expectation formation, including (i) static expectations, which we assumed in Chapters II, III, and IV. In this chapter we examine the further four.

(ii) Myopic perfect foresight, in which expectations are assumed to be fulfilled in the short run. This was stated in equations (5.15) and (5.16),

$$e = g \equiv \dot{v}/v. \tag{5.38}$$

Myopic perfect foresight is a common assumption in optimal growth models, and is, of course, slightly weaker than perfect foresight.

(iii) Simple adaptive expectations, in which expectations are revised upward if the present rate of change of price exceeds that which had been expected for the present, and vice versa. This was formulated in equation (5.17)

$$\dot{e} = \mu \cdot (g - e), \qquad \mu > 0. \tag{5.39}$$

(iv) The expectation of price level formulation, in which resource owners ignore their possible capital gains and consider not the rate of change of price, but the price level. From the adaptive equation (5.26)

$$\dot{v}^e = \alpha \cdot (v - \bar{v}), \tag{5.40}$$

where $\alpha < 0$ ($\alpha > 0$) corresponds to regressive (progressive) expectations, and where the "normal" price $\bar{v}$ is a weighted average of past prices, given by equation (5.21) with β a positive constant, we obtained the formulation in terms of rates of change (equation (5.28))

(5.41) $\dot{e} = \alpha g - ge - \beta e.$

We consider this formulation to see the consequences of such behaviour, even if resource owners have more sophisticated knowledge of their portfolios.

(v) Compound adaptive expectations, in which expectation of the rate of change of price is affected by a belief that the economy will tend to a state in which the asset/resource stock market is in equilibrium and the Hotelling principle holds:

(5.42) $\dot{e} = \gamma\cdot(r-g) + \mu\cdot(g-e),$ $\mu > 0,$

where $\gamma > 0$ ($\gamma < 0$) corresponds to regressive (progressive) expectations, from equation (5.30). In a closely related mode, equation (5.86), the interest rate is compared, not with the actual, but with the expected rate of change of resource price.

We shall see in section 5.4.2 that the assumption of myopic perfect foresight cannot accommodate disequilibrium behaviour, with unfulfilled expectations. In sections 5.5.2 and 5.6.2 we see that, with simple adaptive expectations or with price level expectations, all three possible quasi-equilibria are locally stable if resource supply is completely expectation-inelastic at quasi-equilibrium. Compound adaptive expectations and the direct expectational formulation (sections 5.7 and 5.8) are found to result in local stability under certain conditions, but instability is certain (with the possibility of the Hotelling principle's holding) only in a special case (equation 5.107)). There is no definite tendency for the system to be less stable if the resource owners have progressive expectations.

5.4. Myopic perfect foresight.

5.4.1. The simple resource supply function.

This is a case of the simple resource flow supply function and myopic perfect foresight. The system can be described formally as

$$\dot{v}^e/v \equiv e = g, \tag{5.43}$$

$$\dot{v}/v \equiv g = \lambda_V R^X - \lambda_P Y^X,$$

$$\dot{w}/w \equiv h = \lambda_W N^X - \lambda_P Y^X,$$

$$R^S = R^{SS}(e).$$

Consider Figures 4.3 and 4.4: as long as $R^{SS}(0)$ is positive, equilibrium with constant real prices and fulfilled expectations is unchanged from the inelastic case with $\bar{R}^S = R^{SS}(0)$, since the constant-v locus, and hence the quasi-equilibrium point, is unchanged. This means that in this model, as long as some resource flow is available at an expected rate of change of resource price of zero, there exists a quasi-equilibrium, with zero rate of change of real wage and real resource price, but in general with non-market-clearing and non-zero rates of change of nominal prices of output, labour, and resource.

The stability of such quasi-equilibria, however, will have changed. For the moment we assume that nominal wage W and nominal price of output P are both constant. The model reduces to

$$e = g = \lambda_V(R^D(v) - R^{SS}(e)). \tag{5.44}$$

Consider the plot of e against v. It is necessary for stability of the system that at its v-intercept the $e = \dot{v}/v$ line is negatively sloped, so that for v less than the intercept value v is increasing, and for v greater than the intercept value v is decreasing. This is shown in Figure 5.3.

Differentiating and manipulating the equation above, we get

$$(5.45) \qquad de/dv = \lambda_V R_V^D/(1+\lambda_V R_e^{SS})$$

$$\lesseqgtr 0 \quad \text{as} \quad -R_e^{SS} \lesseqgtr 1/\lambda_V.$$

This can be interpreted as stating that a necessary condition for stability (with constant W and P) is that the absolute value of the slope of the resource supply function is less than the reciprocal of the speed of adjustment of the resource. The higher the speed of adjustment, the less likely this is to hold.

Along the e = g line of myopic perfect foresight, we can evaluate the price elasticity of resource flow fupply:

$$(5.46) \qquad dR^{SS}/dv = R_e^{SS}\cdot de/dv.$$

We can return to the larger model of equations (5.43) including changing nominal wage and nominal price of output as well as nominal resource flow price, and consider the stability of the three possible quasi-equilibria resulting from the simple price adjustment formulation:

(i) Q^{DC}, shown in Figure 4.4a. Analysis of perturbations from this point in Appendix D1 shows that this quasi-equilibrium if it exists, is stable if

$$(5.47) \qquad dg/dv < 0$$

and if, respectively,

$$(5.48) \qquad dg/dw \lesseqgtr 0 \qquad \text{and} \qquad dh/dv \gtreqless 0.$$

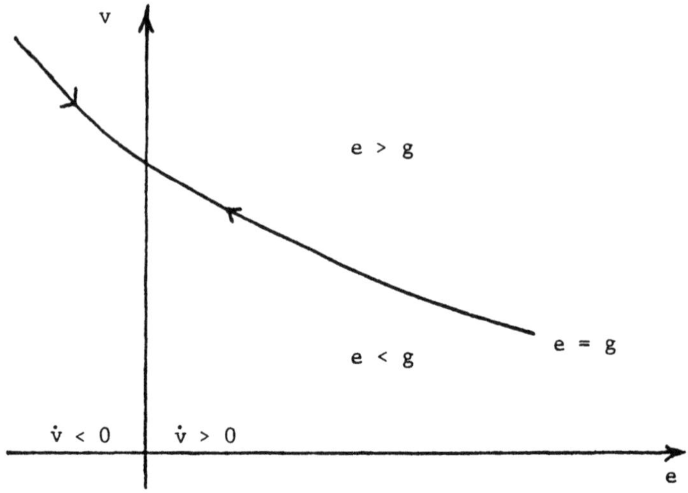

(a) stable

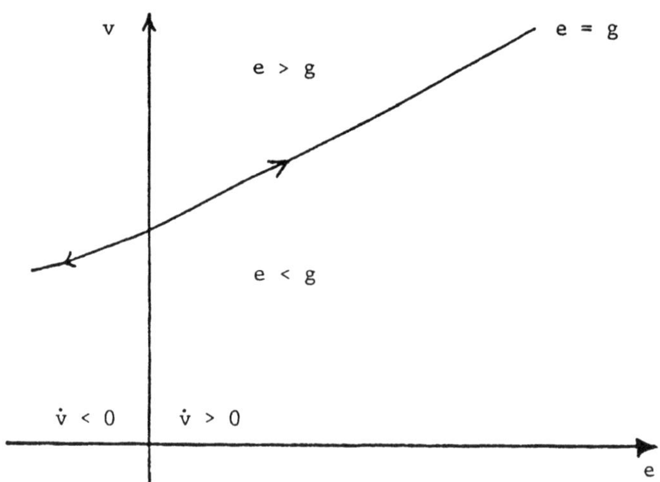

(b) unstable

Figure 5.3: Stability of myopic perfect foresight with $\dot{W} = \dot{P} = 0$.

That is, it is sufficient for a stable quasi-equilibrium that

$$(5.49) \qquad \lambda_V < -1/R_e^{SS} > 0$$

together with, respectively,

$$(5.50) \qquad \lambda_V/\lambda_P \lesseqgtr - Y_w^{SCS}/R_w^{DCD} > 0, \quad \text{and}$$

$$\lambda_W/\lambda_P \gtreqless - Y_v^{SCS}/N_v^{DCD} > 0.$$

The economic interpretation of these conditions is not obvious, but with high λ_V or low λ_W/λ_P the equilibrium will be unstable, although high λ_W/λ_P is not sufficient for stability.

(ii) Q^{NRC}, shown in Figure 4.4b. Analysis of this quasi-equilibrium in Appendix D1 shows that it is sufficient for stability that

$$(5.51) \qquad dg/dv < 0,$$

that is, that

$$(5.52) \qquad \lambda_V < - 1/R_e^{SS} + sF_R \lambda_P.$$

So, for low rates of price adjustment in the market for resource flow, the quasi-equilibrium Q^{NRC} is stable. In addition, we show in Appendix D1 that Q^{NRC} is a stable node.

(iii) Equilibrium Q^{DNR}, shown in Figure 4.3, is easily shown to be a stable node, since a branch of the constant-w locus is horizontal through Q^{DNR}, and a branch of the constant-v locus is vertical through Q^{DNR}.

5.4.2. The quasi-Hotelling resource supply function.

The case of the quasi-Hotelling resource supply function and myopic perfect foresight. The system can be described by

$$
\begin{aligned}
(5.53) \qquad e &= g \\
g &= \lambda_V R^X - \lambda_P Y^X \\
h &= \lambda_W N^X - \lambda_P Y^X \\
R^S = R^{HS}(e) &\equiv \begin{cases} R_2^S & , \quad e \leq r - a, \quad a > 0 \\ R_1^S < R_2^S, & e \geq r + a \end{cases} \\
dR^{HS}/de &= -b < 0 \qquad -a < e - r < a.
\end{aligned}
$$

We shall see, in attempting to analyse this system, that the assumption of myopic perfect foresight becomes unworkable with the quasi-Hotelling resource supply function, since myopic perfect foresight says nothing about the adjustment if expectations are not fulfilled. This problem will be resolved by the inclusion of an adaptive expectations formulation in the system, to include imperfect foresight.

What can be said about the behaviour of an economic system with a quasi-Hotelling resource supply function? From the graph of this function, Figure 5.4a, we shall derive a plot of g against v, assuming myopic perfect foresight, and for the moment ignoring the effects of changes in the real wage by assuming that P and W are constant.

From the first two equations we obtain

$$e = g = \lambda_V(R^D - R^S). \tag{5.54}$$

For any fixed real wage and real resource price, an increase in the flow supply of natural resource will decrease the proportional growth of v, and so in Figure 5.4b the two curves of e against v are drawn with $R_1^S < R_2^S$. The curves have been drawn assuming

$$\lambda_V < -1/R_e^{HS}. \tag{5.55}$$

Consider a value of $e > r + a$. From Figure 5.4a this must coincide with a supply R_1^S. Since $g = e > 0$, in Figure 5.4b the economy will move to the left along the R_1^S path with decreasing expectations, until $e = r + a$ at A. Since $g = e > 0$ at A, the economy cannot move left from A to B: to do so would require $g \equiv \dot{v}/v < 0$.

Consider a value of $e < r - a$. From Figure 5.4a this must coincide with a supply R_2^S. Since $g = e > 0$, in Figure 5.4b the economy will move to the right along the R_2^S path with decreasing expectations, until $e = g = 0$ and the stable quasi-equilibrium of Q is reached. If expectations are such that $-a < e - r < a$, then there is a changing supply of resource flow. Since $g = e > 0$ the economy will move to the right in Figure 5.4b, along the path from B to A, until A is reached: $g = e$ will increase from $(r - a)$ to $(r + a)$.

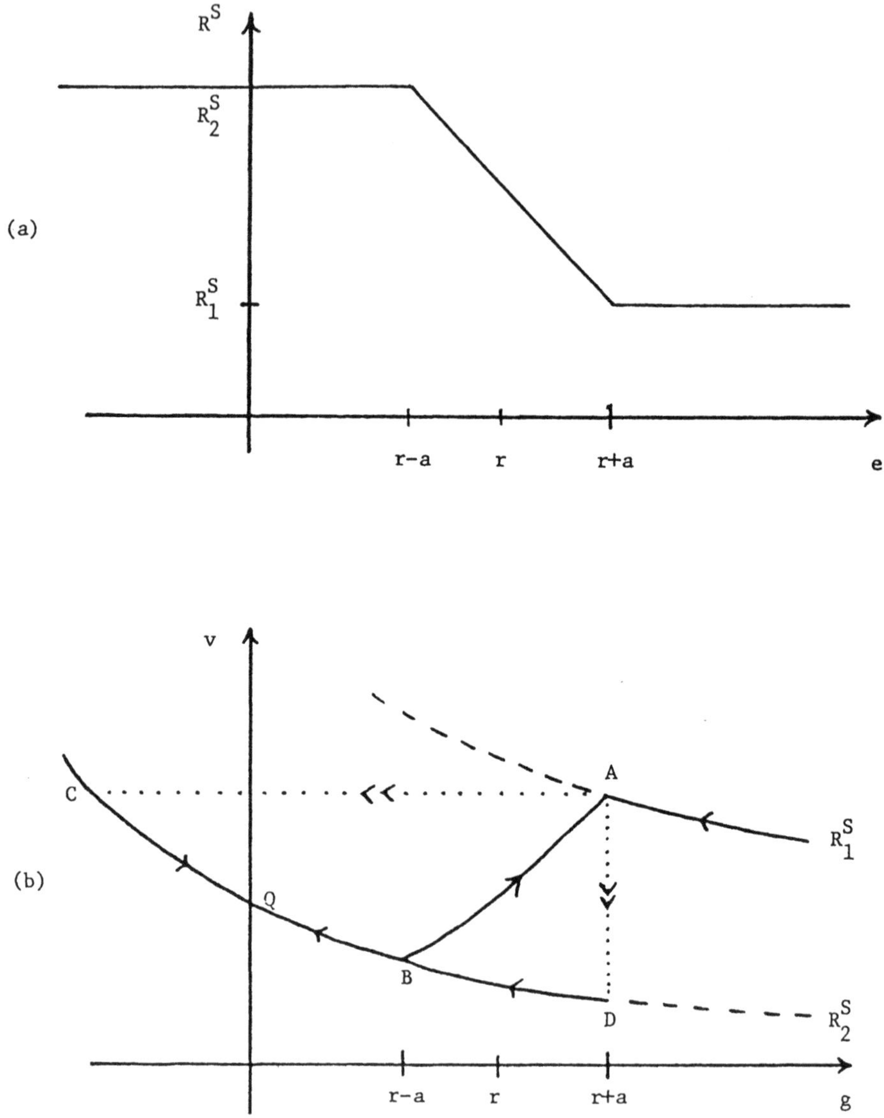

Figure 5.4: The quasi-Hotelling supply function with myopic perfect foresight.

The economy cannot remain at A: since $g = e > 0$, v must increase, but to do so along the R_1^S path would defy the supply function in Figure 5.4a. Any other movement, for example a jump from A to C or to D or to somewhere on the R_2^S path between C and D, would require a discontinuous change in expectations or in resource price or in both. The system as described does not include such behaviour.

Resolution to this dilemma--how does the system behave at point A in Figure 5.4b?--can be resolved by examining what the behaviour of expectations would be if not fulfilled, that is, how will e change if $e \neq g$? So far the assumption has been made that $e = g$, that expectations are fulfilled: with smooth functions, once achieved, this equality may be maintained, but with the discontinuous functions that occur when a quasi-Hotelling resource flow supply function is considered, we should consider the possibility of $e \neq g$, and postulate an adjustment process, based on our understanding of people's modes of forming expectations.

The simple adaptive expectations formulation is written as

$$(5.56) \qquad \dot{e} = \mu(g - e), \qquad \mu > 0.$$

It implies that if $e \lesseqgtr g$, then $\dot{e} \gtreqless 0$, and $e \to g$, that is, it models behaviour whereby if the expected rate of change of price at any moment is not fulfilled, the expectation changes so as to approach the observed behaviour. One way of thinking about the previous case of myopic perfect foresight is that the constant μ is unbounded above, so that $e = g$. But if $\dot{g}$ is bounded and $e = g$ at any time, there is no need to postulate unbounded μ in order that $e = g$ at all succeeding instants. The quasi-Hotelling formulation introduces the possibility of discontinuous shifts in g.

We shall return to the quasi-Hotelling resource flow supply function after first examining the simple resource flow supply function with adaptive expectations.

5.5. Simple adaptive expectations.

5.5.1. The simple resource supply function.

The case of simple resource flow supply function and simple adaptive expectations. The system can be described by

$$(5.57)\qquad \dot{e} = \mu(g - e), \qquad \mu > 0$$

$$g = \lambda_V R^X - \lambda_P Y^X$$

$$h = \lambda_W N^X - \lambda_P Y^X$$

$$R^S = R^{SS}(e).$$

With simple adaptive expectations, it is not possible to equate e with g to obtain the relationship of g versus v, as was done in the cases above. Instead we have added a dimension to the system: we now have three state variables: w, v, and e.

If the adaptive expectations formulation is written in discrete form, we get

$$(5.58)\qquad e(t) - e(t-1) = \mu'.(g(t-1) - e(t-1)), \quad 0 \le \mu' \le 1.$$

After manipulation, this becomes

$$(5.59)\qquad e(t) = \mu'.g(t-1) + (1 - \mu').e(t-1).$$

With $\mu' = 1$,

(5.60) $$e(t) = g(t-1)$$

which states that the expected proportional rate of change of price in a period equals the actual rate of change in the previous period. If $0 < \mu' < 1$, then the new expectation is a weighted average of the old expectation and the actual value. As the periods become even shorter, with $\mu' = 1$, this case approaches that of myopic perfect foresight discussed above, $e = g$.

Using the discrete formulation, we examine the system above, assuming for the moment that nominal wage W and nominal price of output P are unchanging. The system becomes

(5.61) $$e(t) = \mu'.g(t-1) + (1-\mu').e(t-1), \qquad 0 < \mu' < 1$$

$$g(t) = \lambda_V(R^D(v(t)) - R^{SS}(e(t)))$$

$$v(t) = v(t-1).(1+g(t-1)).$$

The simple, downwards-sloping resource flow supply function of Figure 5.2 is redrawn in Figure 5.5a. Using this graph, we shall derive a plot of expected rate of change of price e against actual price v, with the intermediate plot of the actual rate of change of price g as a family of functions of the expected rate of change e, for different values of the price v.

From the expression for actual rate of change, g, we get

(5.62) $$\frac{\partial g}{\partial e} = -\lambda_V R_e^{SS} > 0 \quad \text{since } R_e^{SS} < 0$$

$$\frac{\partial g}{\partial v} = \lambda_V R_v^D \leq 0 \quad \text{as } R_v^D \leq 0.$$

Figure 5.5b has been drawn with $R_v^D < 0$, with $\lambda_V < -1/R_e^{SS}$, and with $v_1 < v_2 < v_3$. Each of the curves $g = g(v_i)$ shows the actual rate of change against the expected rate of change for a given value of the price. The higher the price, the lower the curve, from the consequently lower excess demand for resource flow. The 45-degree line is the locus of points for which expectations are fulfilled at any instant, $e = g$. Note that only if $e = g = 0$ can we speak of unchanging, fulfilled expectations, since if $g \neq 0$, the g-curve will be shifting as the price changes. The price consistent with zero actual and expected rate of change of price in Figure 5.5b is v_3, and, as drawn, this is a stable equilibrium: $g \lesseqgtr 0$ as $v \gtreqless v_3$. It will be seen from the graph that it is necessary for stability that $\partial g/\partial e < 1$, otherwise, as v increases for positive g, the point of fulfilled expectations $e = g$ will move to the NE along the 45-degree line, and $g \to \infty$. This is a graphical exposition of the stability condition derived algebraically in section 5.4.1 above as equation (5.45),

$$(5.63) \qquad \lambda_V < -1/R_e^{SS}.$$

At any time $t = 1$ there will be a price $v(t)$, say v_1. Let the expectation $e(t)$ at time $t = 1$ be e_1. From these two values the demand and supply, respectively, are determined, and from the state of the resource flow market the actual rate of change of price, $g(t)$, say g_1. On Figure 5.5b, g_1 is determined by the intersection of the vertical line e_1 with the price change line corresponding to v_1, as shown. If this point lies on the $e = g$ 45-degree line, then the momentary expectation e_1 is fulfilled at time $t = 1$, since this expectation leads to a resource flow supply which, with the demand from price v_1, leads to an actual

(a)

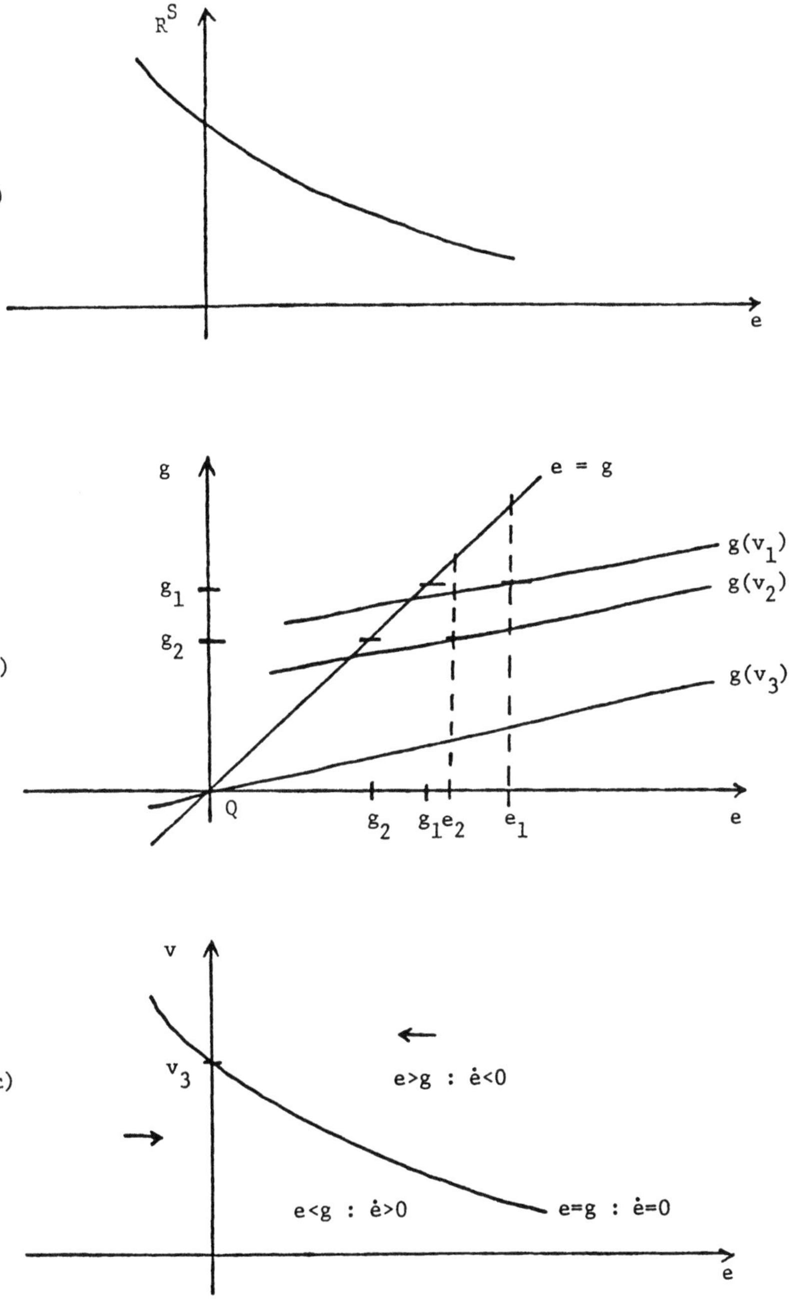

(b)

(c)

Figure 5.5: The simple supply function with simple adaptive expectations.

change of price g_1 which equals the expected rate of change e_1. Note, however, that as long as $g \neq 0$, the system is not in quasi-equilibrium since by period $t = 2$ the price, and with it the rate of change of price, will have changed, and so the expected rate of change (if unchanged) will not be realized.

In Figure 5.5b, as drawn, $e_1 > g_1$, leading to a revision of expected rate of change of price downward so that in period $t = 2$ we have e_2: $g_1 < e_2 < e_1$. Meanwhile, of course, the actual price level will have changed to v_2, which together with e_2 leads to the excess demand in the resource flow market, and hence the actual rate of change of price g_2. As mentioned above, for the family of curves $g(v)$ having a slope less than one, which is equivalent to the condition that

(5.64) $$\lambda_V < -1/R_e^{SS}.$$

Figure 5.5b shows that $g_2 < g_1$, with the (stable) end result of this process, the quasi-equilibrium $g = e = 0$ at v_3, as drawn. The rate of convergence to this point cannot easily be deduced from the figure.

From Figure 5.5b we can derive a plot of the expected rate of change of price e against the level of price v, as shown in Figure 5.5c. This is achieved by plotting for each v the point of intersection of the $g(v)$ curve with the 45-degree fulfilled expectation line, $e = g$. This gives the locus of points along which expectations are fulfilled, which corresponds to myopic perfect foresight. The locus is downwards sloping, which means that the equilibrium at v_3 is stable for myopic perfect foresight:

(5.65) $$v \gtrless v_3 \quad \text{iff} \quad g \lesseqgtr 0.$$

From Figure 5.5b, for a given level of expected rate of price change e, as the price v increases (decreases), the actual rate of price change g falls (rises), which means that to the NE (SW) of the fulfilled expectations locus $e = g$ in Figure 5.5c, expected rate of change is greater (lesser) then actual $e > g$ ($e < g$), which results in a falling (rising) level of expectations $\dot{e} < 0$ ($\dot{e} > 0$), as indicated by the arrows.

Along the fulfilled expectations locus, $\dot{e} = 0$: the fulfilled expectations locus with adaptive expectations is the locus of constant expected rate of price change. By plotting the locus of constant price $\dot{v} = 0$, we can complete a phase diagram for the simple system, which in continuous formulation is

$$(5.66) \qquad \dot{e} = \mu(g - e)$$

$$g = \lambda_V (R^D(v) - R^{SS}(e)).$$

The two Figures 5.6a and 5.6b have been drawn for the conditions, respectively

$$(5.67) \qquad -R_e^{SS} \lesseqgtr 1/\lambda_V.$$

It is not possible for the constant-e locus to be steeper than the constant-v locus since

$$(5.68) \qquad \left.\frac{dv}{de}\right|_{\dot{e}} = \left.\frac{dv}{de}\right|_{\dot{v}} + 1/\lambda_V \, R_v^D, \qquad \text{with } R_v^D \le 0.$$

Analysis in Appendix D2 shows that a necessary and sufficient condition for stability of the system of equations (5.66) is that

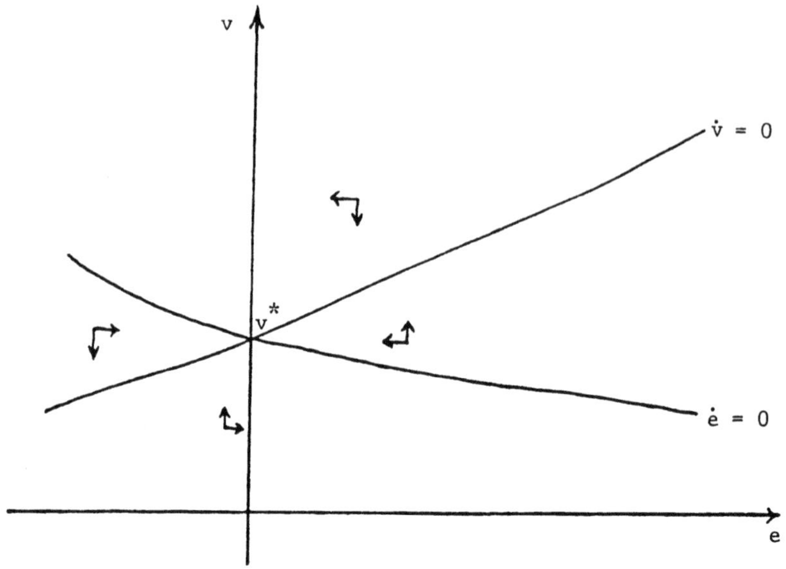

(a) $\lambda_V < -1/R_e^{SS}$

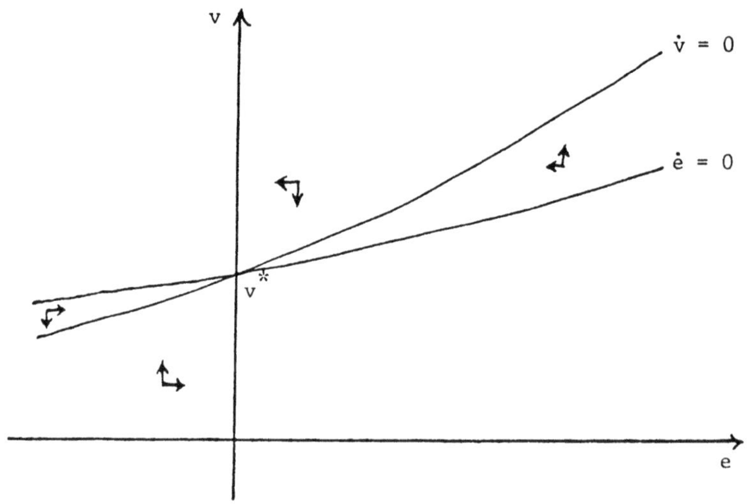

(b) $\lambda_V > -1/R_e^{SS}$

Figure 5.6: Phase diagrams, simple resource supply, simple adaptive expectations.

$$(5.69) \qquad -R_e^{SS} < 1/\lambda_V - v^* R_v^D/\mu$$

where v^*, previously written as v_3, is the resource flow price at which $e = g = 0$. This condition is equivalent to

$$(5.70) \qquad \left.\frac{dv}{de}\right|_{\dot{e}} < v^*/\mu.$$

This condition means that it is not necessary for stability that

$$(5.71) \qquad \lambda_V < -1/R_e^{SS}$$

as previously claimed, since $R_v^D \le 0$.

Allowing nominal wage W and nominal price of output P to change in response to excess effective demands leads to the system's having three state variables: w, v, and e in equations (5.57). The phase diagram, being three-dimensional, cannot be easily drawn.

Analysis in Appendix D2 for the quasi-equilibria Q^{DC} in the region of Keynesian unemployment, and Q^{NRC} in the region of repressed inflation, shows that it is sufficient for local dynamic stability of the first that inequality (5.64) hold (see inequalities (5.52) and (4.19))

$$(5.72) \qquad \lambda_V < -1/R_e^{SS},$$

and that it is sufficient for local dynamic stability of the second that the two inequalities below hold

$$(5.73) \qquad \lambda_V R_e^{SS} - s\,\lambda_P F_R R_e^{SS} + 1 > 0,$$

$$\lambda_V/\lambda_P < s\,F_R.$$

The second states that the ratio of the speeds of adjustment of resource and output money prices should be less than the fraction of the marginal product of resource saved.

5.5.2. The quasi-Hotelling resource supply function.

The case of the quasi-Hotelling resource flow supply function and adaptive expectations. The system can be described by

$$(5.74) \qquad \dot{e} = \mu(g - e), \qquad \mu > 0,$$

$$g = \lambda_V R^X - \lambda_P Y^X,$$

$$h = \lambda_W N^X - \lambda_P Y^X,$$

$$R^S = R^{HS}(e).$$

By using a discrete formulation of this system, and assuming for the moment that nominal wage W and nominal price of output P are unchanging, we can derive a plot of actual price v against expected rate of change of price e, in a manner similar to that used in Figure 5.5 for the simple supply function. The simplified system becomes

$$(5.75) \qquad e(t) = \mu'\cdot g(t-1) + (1-\mu')\cdot e(t-1), \qquad 0 < \mu' < 1$$

$$g(t) = \lambda_V(R^D(v(t)) - R^{HS}(e(t)))$$

$$v(t) = v(t-1)\cdot(1 + g(t-1)).$$

The family of curves $g = g(v_i)$, each of which shows the actual rate of change of price g against the expected rate of change e for a given price level v_i have been plotted in Figure 5.7b for $R_v^D < 0$. Partial differentiation of actual rate of change g gives

$$(5.76) \qquad \frac{\partial g}{\partial e} = -\lambda_V R_e^{HS} = \begin{cases} 0, & -a \geq e-r \geq a \\ \lambda_V b, & -a < e-r < a \end{cases}$$

$$\frac{\partial g}{\partial v} = \lambda_V R_v^D < 0 \quad \text{as} \quad R_v^D < 0.$$

The diagram has been drawn with $\lambda_V > 1/b$, i.e. $\lambda_V > -1/R_e^S$, which means that we shall obtain a downwards-sloping segment of the constant-e locus, corresponding to segment AB in Figure 5.4b. It is the existence of this segment which led us to realise that the assumption of myopic perfect foresight was inadequate to describe the economy. The graph is drawn with $v_1 < v_2 < v_3 < v_4$; v_4 is the price consistent with zero actual and expected rate of change of price, although it is possible, if the change in supply behaviour in the region $-a < e-r < a$ is rapid enough, i.e. if b is large enough, that there are three levels of actual rate of change of price (one zero, two positive) consistent with v_3 and momentary fulfilled expectations (e=g), as seen in Figure 5.4b.

At any time t=1 there will be a price v(t), say v_1. This will result in a level of demand for resource flow from industry. At any time t=1 there will be a level of price expectation e(t), say e_1, which will result in a level of supply of resource flow from the holders of resource stocks. The rate of change of price will respond to the excess demand in the resource flow market, which will result in actual rate of change of price, g(t), say g_1. In Figure 5.7b, g_1 is determined by the point on $g(v_1)$ corresponding to e_1. If this point lies on the fulfilled expectations line e=g, then the momentary expectation e_1 is fulfilled: $e_1 = g_1 = g(v_1)$; but as long as $g \neq 0$ the system is not in quasi-equilibrium since by period t=2 the price, and with it the actual rate of change of price, will have changed, leading to unfulfilled expectations in period t=2.

In Figure 5.7b, as drawn, $e_1 > g_1$, leading to a downward revision of expected rate of price change so that in period t=2 we have e_2: $g_1 < e_2 < e_1$. Note that it is possible to have an upward revision of

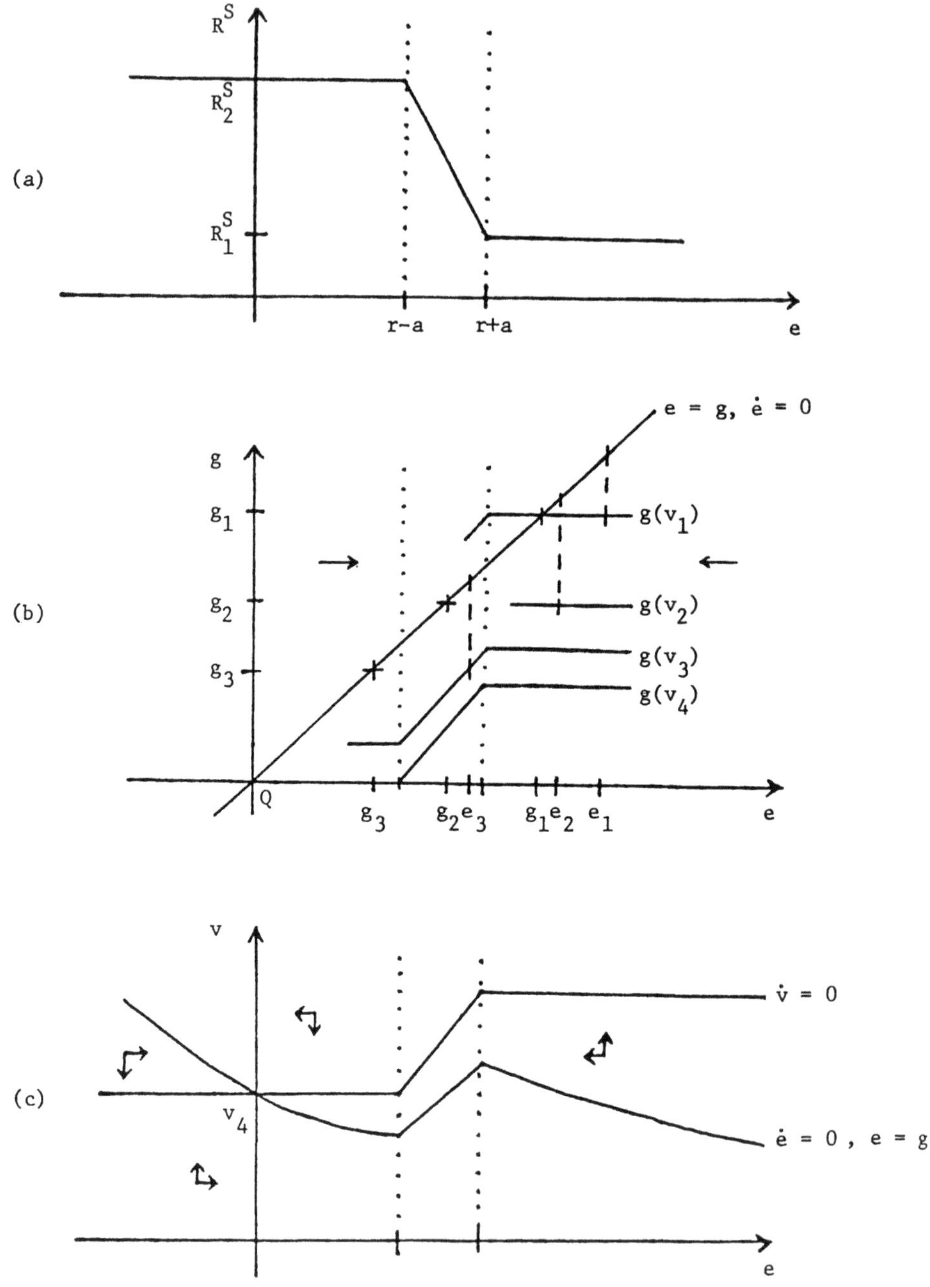

Figure 5.7: The quasi-Hotelling supply function with simple adaptive expectations.

expected rate of price change if the actual is greater than the expected rate at any instant; thus if $e = e_3$ with $v = v_1$, then $e < g$ in this period and expectations will increase. This is unlikely to persist for long, however, since rising price v will lead to a falling actual rate of price rise, which will eventually be less than the expected rate.

At time t=2, since $g_1 > 0$, the actual price v(t) will have increased, leading to $v(t) = v_2$; the expected rate of change of prices, as drawn, will have fallen to e_2, and together these two parameters, through the resource flow market, will determine the actual rate of price change g_2. Since $e_2 > g_2$, there will be a further downwards revision of expectations, which at time t=3 will be e_3. As drawn, $e_3 < r+a$, which means that the amount of resource for sale on the flow market is suddenly reduced as holders see the expected capital gain from holding other assets (as given by the interest rate r) becoming more attractive than the expected capital gain from holding stocks of natural resource (as given by the expected rate of price rise $e_3 < r+a$). This reduction in R^S leads to a reduction in excess demand and a sudden drop in the actual rate of change of price, to $g = g(v_3) = g_3$.

Expectations will be revised downwards and price will rise. The end result of this process will be zero actual and expected rate of change of price, with price equal to v_4. There may be some time during which the price and its actual rate of change are constant at v_4 and zero respectively, while the expected rate of price change falls to zero: for $e < r-a$ there is no change in the flow supply of resource, R_2^S.

So analysis of a simple model with a quasi-Hotelling resource flow supply function and adaptive expectations shows that the process converges to the point of zero actual and expected proportional rate of change of price, with no great jumps of either the price or its actual rate of change, as had been postulated as resolution to the process with myopic perfect foresight in section 5.4.2 above. If $\lambda_V < 1/b$, the reader can easily demonstrate that the process will be substantially similar to that described above, with a less rapid approach to the quasi-equilibrium at the origin.

From Figure 5.7b we can derive a plot of the price v against its expected rate of change e, for momentarily fulfilled expectations. This is done by plotting for each v the point of intersection of the g(v) curve with the fulfilled expectation line, e=g. As mentioned above, this may not be unique: there exists a range of prices for which there are three expectations of rate of price change which are fulfilled. The locus of momentarily fulfilled expectations, e=g, is shown in Figure 5.7c. Through the adaptive expectation equation, it corresponds to the constant-e locus, $\dot{e}=0$. Addition of the constant-v locus, g=0, will result in a phase diagram, for the case of W and P constant. In figure 5.7c the constant-v locus has been plotted using the second of equations (5.75): the locus is horizontal except where R^{HS} decreases with e, and to maintain g = 0 the real wage must rise to reduce the demand R^D. From this two-dimensional phase diagram we can see that adjustment will tend to move in a counter-clockwise fashion. Figure 5.7c can be compared with Figure 5.4b, which first indicated the inadequacy of the assumption of myopic perfect foresight in the quasi-Hotelling resource supply case.

Analysis in Appendix D3 of the three-dimensional system, equations (5.74), for the quasi-equilibria Q^{DC} in the region of Keynesian unemployment, and Q^{NRC} in the region of repressed inflation, shows that both are locally stable. Q^{NRC} is a stable node, and, if the production function is Cobb-Douglas, so is Q^{DC}. That these results are identical with those derived in Appendix C3 and stated in section 4.2.1 for the case of the simple formulation of price adjustment and inelastic factor supplies should not surprise us, since in the region of the quasi-equilibrium, shown in Figure 5.7c, the expected rate of change of price of resource is zero, and, from Figure 5.2, the supply of resource flow is inelastic at R_2^S.

5.6. Expectation of price level.

In section 5.2.2 we considered the expectation formulation that would result from resource owners' ignoring the possible capital gains that would accrue from rising v, and instead expecting the real resource price to fluctuate around a "normal" price $\bar{v}$, a weighted average of past prices given by equation (5.21)

$$(5.77) \qquad \bar{v}(t) \equiv \beta\int_{-\infty}^{t} v(x)\cdot \exp^{-\beta\cdot(t-x)}\, dx, \qquad \beta > 0.$$

The adoptive expectations equation in terms of price levels is given by equation (5.26)

$$(5.78) \qquad \dot{v}^e = \alpha.(v - \bar{v}), \qquad \alpha \gtrless 0,$$

where $\alpha < 0$ ($\alpha > 0$) corresponds to regressive (progressive) expectations. Differentiation and manipulation lead to a formulation in terms of rates of change (equation (5.28))

$$(5.79) \qquad \dot{e} = \alpha g - ge - \beta e = (\alpha - e)g - \beta e.$$

The locus of fulfilled expectations ($\dot{e} = 0$) resulting from this equation is plotted in the (e, g)-plane in Figure 5.8: Figure 5.8a is the case of progressive expectations ($\alpha > 0$), and Figure 5.8b the case of regressive expectations ($\alpha < 0$). In each case there are seen to be two branches of the locus, one (through the origin) stable, and the other unstable, as shown by the small arrows.

5.6.1. The simple resource supply function.

The case of the simple resource flow supply function and the expectation of price level formulation. The system can be described by

$$\text{(5.80)} \qquad \begin{aligned} \dot{e} &= (\alpha - e)g - \beta e, \\ g &= \lambda_V R^X - \lambda_P Y^X, \\ h &= \lambda_W N^X - \lambda_P Y^X, \\ R^S &= R^{SS}(e). \end{aligned}$$

This is a three-dimensional system, but for the moment we assume that P and W are unchanging. Figure 5.9 shows the system (equation (5.81)) plotted in the (e, v)-plane: Figure 5.9a is the case of progressive expectations ($\alpha > 0$), and Figure 5.9b the case of regressive expectations ($\alpha < 0$). We see that in both cases quasi-equilibrium occurs with expectations of constant v ($e = 0$) and positive v.

The simple, two-dimensional system can be described as

$$\text{(5.81)} \qquad \begin{aligned} \dot{e} &= (\alpha - e)g - \beta e, \\ g &= \lambda_V(R^D(v) - R^{SS}(e)). \end{aligned}$$

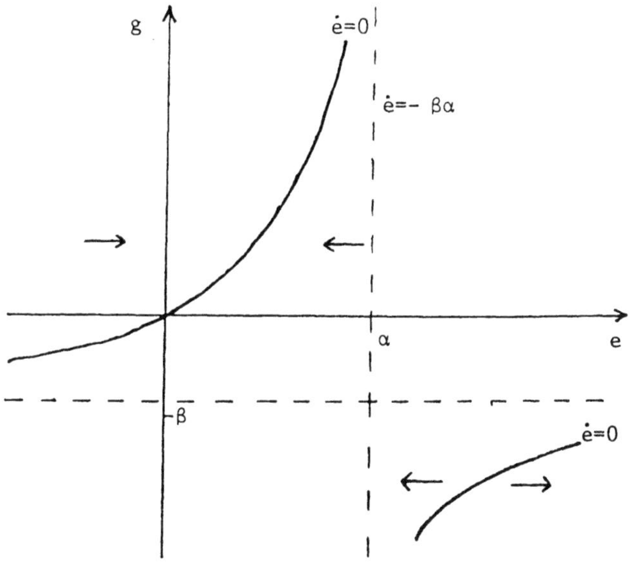

(a) α>0 , progressive expectations

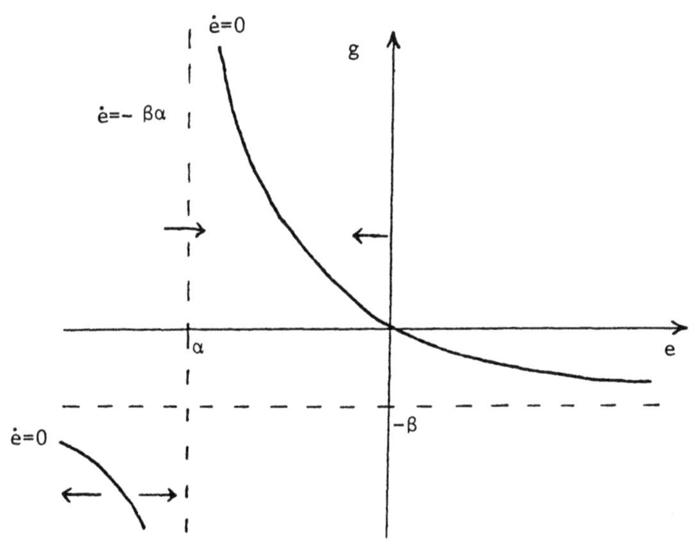

(b) α<0 , regressive expectations

Figure 5.8: <u>The fulfilled expectations locus, price level expectations.</u>

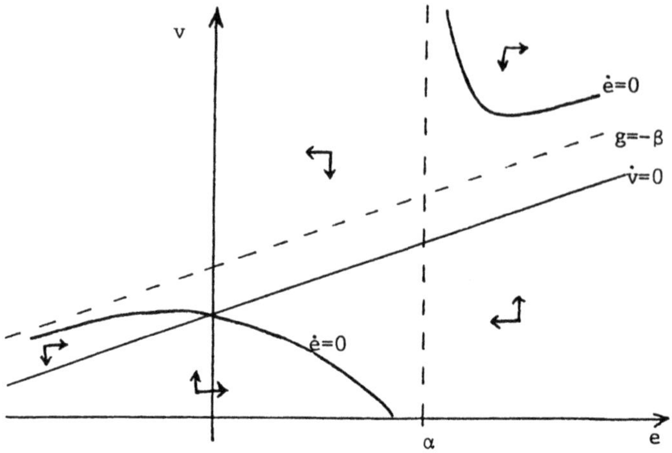

(a) α>0, progressive expectations

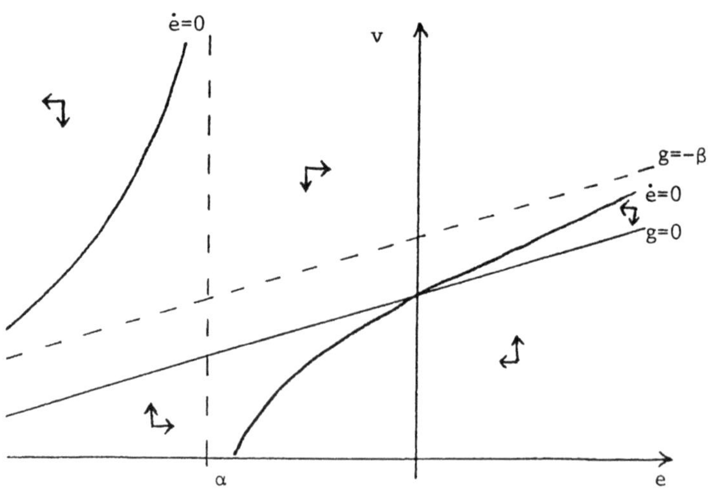

(b) α<0, regressive expectations

Figure 5.9: Phase diagrams, simple resource supply, price level expectations.

Analysis in Appendix D4 shows that this system is locally dynamically stable iff inequality (D49) holds

$$(5.82) \qquad -\alpha R_e^{SS} < \beta/\lambda_V - v^* R_v^D.$$

With progressive expectations ($\alpha > 0$) this condition imposes a limit on the slope of $R^{SS}(e)$; with regressive expectations ($\alpha < 0$) there is no limit.

Allowing money wage W and money price of output P to vary in response to excess effective demands in the markets for labour and output respectively leads to a three-dimensional system. Analysis in Appendix D4, for the quasi-equilibria Q^{DC}, in the region of Keynesian unemployment, and Q^{NRC}, in the region of repressed inflation, shows that it is sufficient for local dynamic stability of the first that expectations are regressive ($\alpha < 0$), and of the second that the expectations are progressive ($\alpha > 0$) if, from inequality (D60),

$$(5.83) \qquad \lambda_V/\lambda_P < s\ F_R,$$

or regressive ($\alpha < 0$) otherwise. If expectations are progressive with large α, then we argue in Appendix D4 from inequality (D52) that Q^{DC} is unstable.

5.6.2. The quasi-Hotelling resource supply function.

The case of the quasi-Hotelling resource flow supply function and the expectation of price level formulation. The system can be described by

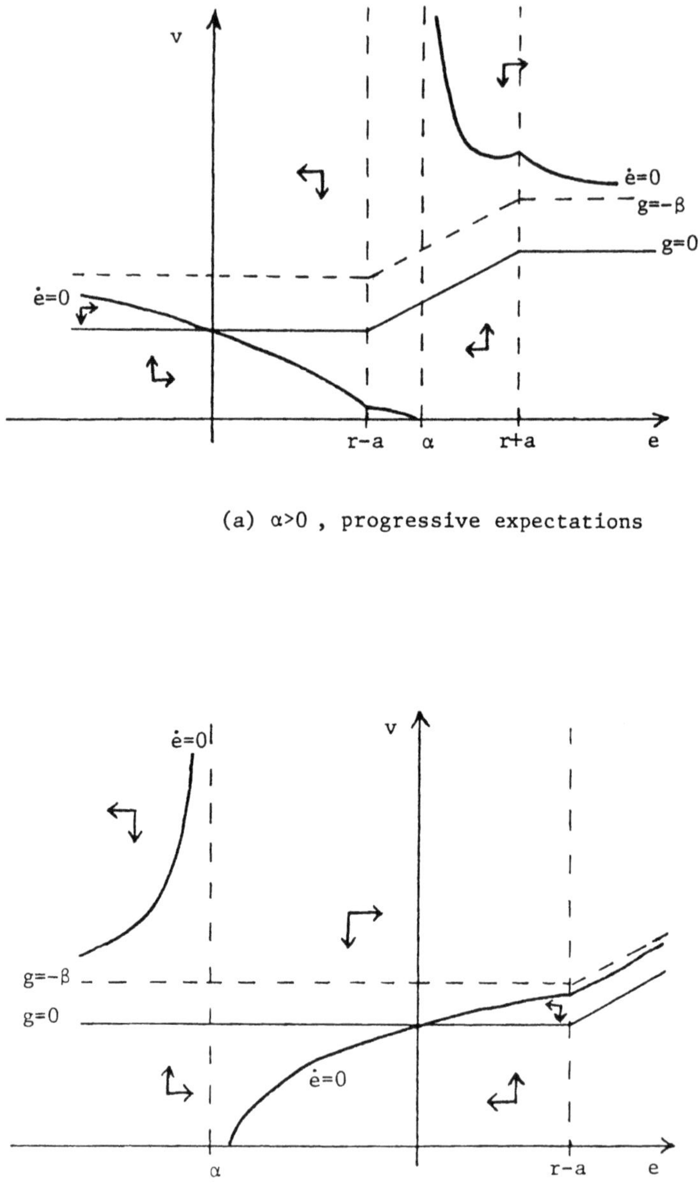

(a) α>0 , progressive expectations

(b) α<0, regressive expectations

Figure 5.10: Phase diagrams, quasi-Hotelling resource supply, price level expectations.

$$(5.84)\qquad \begin{aligned} \dot{e} &= (\alpha - e)g - \beta e, \\ g &= \lambda_V R^X - \lambda_P Y^X, \\ h &= \lambda_W N^X - \lambda_P Y^X, \\ R^S &= R^{HS}(e). \end{aligned}$$

As in section 5.5.2 this is a three-dimensional system, but for the moment we assume that P and W are unchanging. Figure 5.10 shows the system plotted in the (e, v)-plane. Figure 5.10a is the case of progressive expectations ($\alpha > 0$), and Figure 5.10b the case of regressive expectations ($\alpha < 0$). We see that in both cases quasi-equilibrium occurs with positive v, expectations of constant v ($e = 0$), and $R^S = R^{HS}(0) = R_2^S$ from Figure 5.2.

Analysis in Appendix D5 of the three-dimensional system of equations (5.84) for the quasi-equilibria Q^{DC} in the region of Keynesian unemployment, and Q^{NRC} in the region of repressed inflation, shows that both quasi-equilibria are locally dynamically stable for both regressive and progressive expectations ($\alpha \gtrless 0$).

5.7. Compound adaptive expectations.

In section 5.2.2 we considered the formulation of expectations that would result from resource owners' being aware of the Hotelling principle as they considered the capital gains that would accrue as the price of resource v rose, and as they compared the actual rate of change of price g with the rate of return r available from alternative investments. The formulation was given by equation (5.30)

$$(5.85) \qquad \dot{e} = \gamma\cdot(r-g) + \mu\cdot(g-e) = (\mu-\gamma)g - \mu e + \gamma r, \qquad \mu > 0,$$

where $\gamma > 0$ corresponds to regressive expectations, since if the actual rate of change g is less (more) than the interest rate r, the expected rate of change e increases (decreases), ceteris paribus; $\gamma < 0$ corresponds to progressive expectations, since if g is less (more) than r, then e decreases (increases), ceteris paribus, leading to a knife-edge situation.

The formulation of equation (5.85) is very similar to that of

$$(5.86) \qquad \dot{e} = \delta\cdot(r-e) + \nu\cdot(g-e) = \nu g - (\nu+\delta)e + \delta r, \qquad \nu > 0.$$

This second formulation differs from the first in that the interest rate is compared, not with the actual rate of change of resource price, but with the expected rate of change. Since the formulation is attempting to model the process whereby the resource owners change their expected rate of change, the second formulation is more direct, even if the comparison may not take place. The parameter ν is positive, since the expected approaches the actual rate of change as in the simple adaptive formulation. The parameter δ can take either sign: δ positive corresponds to regressive expectations, since if the expected rate of change e is less (more) than the interest rate r, e increases (decreases); δ negative corresponds to progressive expectations, a knife-edge.

The two versions of the compound adaptive expectations formulation (equations (5.85) and (5.86)) turn out to be very similar mathematically: each can be obtained as a linear transformation of the other. By analyzing the first formulation (equation (5.85)) qualitatively we are also analyzing the second (equation (5.86)). Comparison of the two formulations shows that if

$$\nu = \mu - \gamma, \quad \text{and} \quad \delta = \gamma,$$

the two are identical.

The locus of fulfilled expectations ($\dot{e} = 0$) resulting from this formulation is plotted in the expected-actual plane (e, g) in Figure 5.11, for various values of the two parameters μ and γ. We have defined quasi-equilibrium to be constant v and constant w, that is, $g^* = h^* = 0$. In Figure 5.11 if, as we assume, quasi-equilibrium occurs with fulfilled expectations ($\dot{e} = 0$), then the quasi-equilibrium in each case is the intersection of the fulfilled expectations locus with the e-axis, at which point $\dot{e} = 0$ and $g = 0$. Then we see that, unlike the earlier formulations of sections 5.5 and 5.6, this formulation admits of the possibility of quasi-equilibrium with $e^* \neq 0$. Indeed, $e^* = 0$ occurs only when $\gamma = 0$, which is the simple adaptive formulation of section 5.5. The expected rate of change of price at quasi-equilibrium e^* is given by, from equation (5.85),

$$e^* = \gamma r/\mu.$$

With regressive expectations ($\gamma > 0$), e^* is positive, but with progressive expectations ($\gamma < 0$), the expected rate of change of price at quasi-equilibrium e^* is negative: if the formulation is realistic, people expect the price to fall and continue falling despite the fact that it is constant at quasi-equilibrium. Cases (a), (b), (c), and (d) of Figure 5.11 show regressive expectations, cases (e) and (f) show progressive expectations. With the second formulation (equation (5.86)), cases (a) and (b) are infeasible since ν is negative, and we can describe two further cases of progressive expectations: (g) with $\delta < 0 < \nu$ (and $\gamma < \mu < 0$), and (h) with $\delta < \nu = 0$ (and $\gamma = \mu < 0$). Both are infeasible with the first formulation (equation (5.85)) and both are unstable, since μ is negative. Case (g)

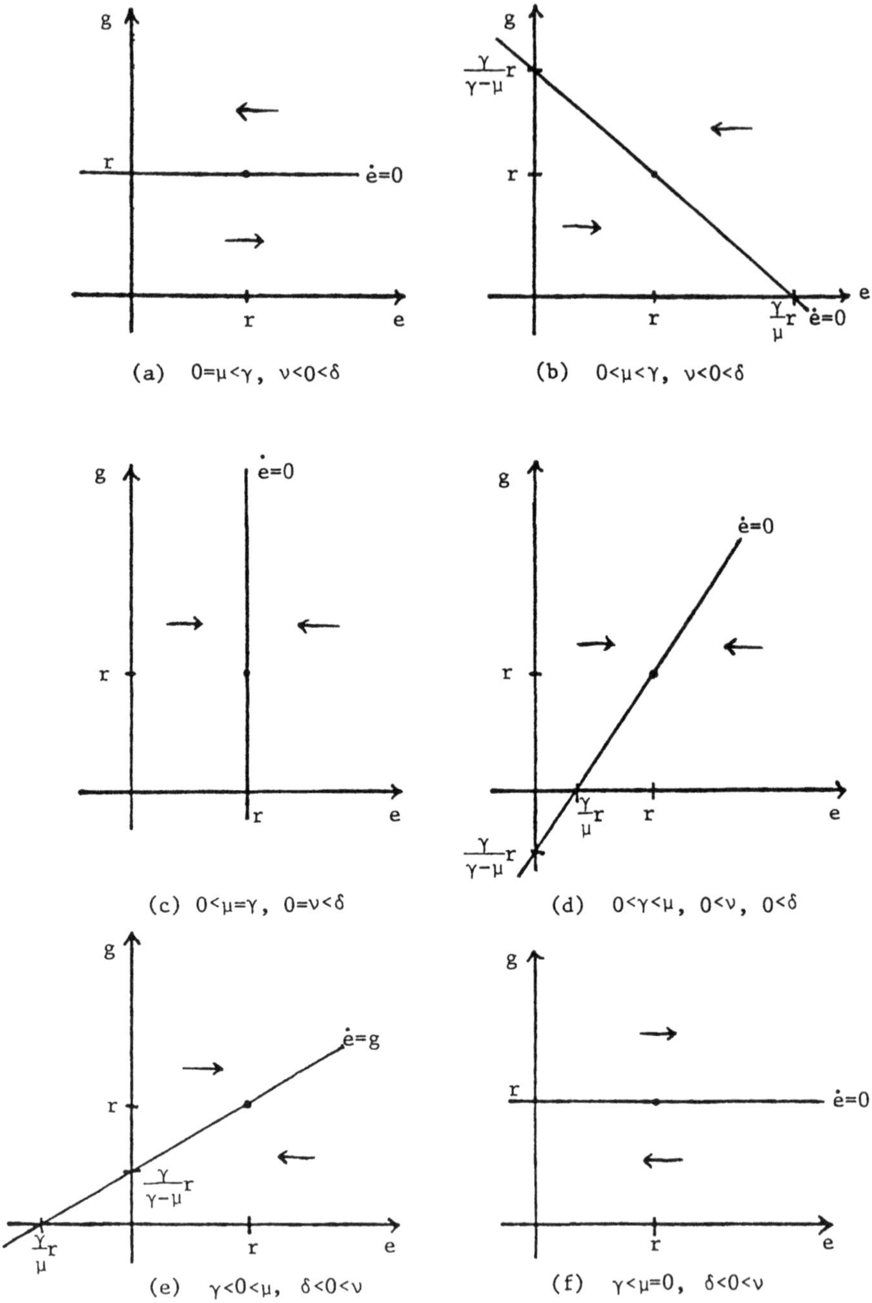

(a) $0=\mu<\gamma$, $\nu<0<\delta$

(b) $0<\mu<\gamma$, $\nu<0<\delta$

(c) $0<\mu=\gamma$, $0=\nu<\delta$

(d) $0<\gamma<\mu$, $0<\nu$, $0<\delta$

(e) $\gamma<0<\mu$, $\delta<0<\nu$

(f) $\gamma<\mu=0$, $\delta<0<\nu$

Figure 5.11: The fulfilled expectations locus, compound adaptive expectations.

resembles case (b), and case (h) resembles case (c), but both with the direction of $\dot{e}$ reversed. From Figure 5.11 and equation (5.85) we see that expectations are always fulfilled ($\dot{e} = 0$) when $e = g = r$: the formulation always allows the Hotelling principle to occur with fulfilled expectations. The movements of adjustment of e are shown in Figure 5.11 by the small arrows.

5.7.1. The simple resource supply function.

The case of the simple resource flow $upply function and the formulation of compound adaptive expectations. The system can be described by

$$(5.87) \qquad \dot{e} = (\mu - \gamma)g - \mu e + \gamma r, \qquad \mu > 0,$$

$$g = \lambda_V R^X - \lambda_P Y^X,$$

$$h = \lambda_W N^X - \lambda_P Y^X,$$

$$R^S = R^{SS}(e).$$

This is a three-dimensional system, but for the moment we assume that P and W are unchanging. Figure 5.12 shows the system (equation (5.88)) plotted in the (e, v)-plane for various values of the parameters μ and γ. Figure 5.12a, with $\mu = 0$ and regressive expectations ($\gamma > 0$), is unstable, with the system moving to the NE between the constant-v and constant-e loci, with $\dot{e}$ positive and less than γr, and g positive and less than r. Figure 5.12f is similarly unstable, except that expectations are progressive and $\gamma < 0$. For $\mu \neq 0$ we see that $e^* = \gamma r/\mu$ is finite: with regressive expectations e^* is positive, with progressive expectations, negative.

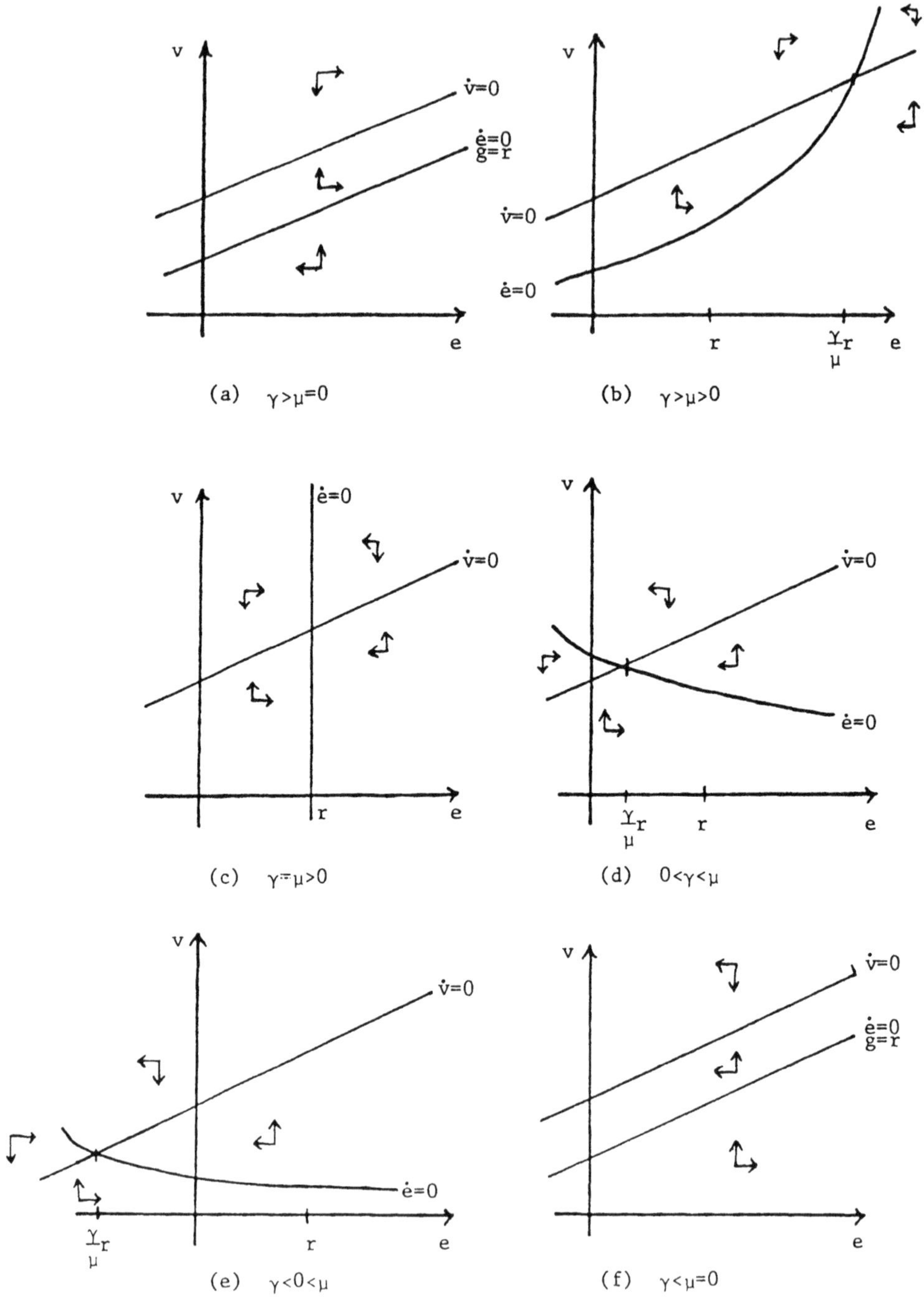

Figure 5.12: Phase diagrams, simple resource supply, compound adaptive expectations.

The two-dimensional system can be written as

$$(5.88) \qquad \dot{e} = (\mu - \gamma)g - \mu e + \gamma r, \qquad \mu > 0,$$

$$g = \lambda_V(R^D(v) - R^{SS}(e)).$$

Analysis in Appendix D6 shows that cases (a), (f), (g), (h) are unstable and that this system is locally dynamically stable iff inequality (D71) holds

$$(5.89) \qquad -(\mu - \gamma)\ R_e^{SS} < \mu/\lambda_V - v^*\ R_v^D.$$

Cases (b) and (c) are shown to be definitely stable; cases (d) and (e) are possibly unstable.

Allowing money wage W and money price of output P to vary in response to excess effective demands in the markets for labour and output respectively leads to the three-dimensional system of equations (5.87). Analysis in Appendix D6, for the quasi-equilibrium Q^{DC}, in the region of Keynesian unemployment, shows that it is definitely stable in cases (b) and (c), definitely unstable in the cases with $\mu \leq 0$, (a), (f), (g), (h), and possible unstable in the cases (d) and (e). The quasi-equilibrium Q^{NRC}, in the region of repressed inflation, is definitely stable in case (c), definitely unstable in cases (a), (f), (g), (h) and possibly unstable in cases (b), (d), and (e).

5.7.2. <u>The quasi-Hotelling resource supply function</u>.

The case of the quasi-Hotelling resource flow supply function and the formulation of compound adaptive expectations. The system can be described by

$$(5.90)\qquad \begin{aligned} \dot{e} &= (\mu - \gamma)g - \mu e + \gamma r, \qquad \mu > 0, \\ g &= \lambda_V R^X - \lambda_P Y^X, \\ h &= \lambda_W N^X - \lambda_P Y^X, \\ R^S &= R^{HS}(e). \end{aligned}$$

This is a three-dimensional system, but for the moment we assume that P and W are unchanging. Figure 5.13 shows the system (equations (5.93)) plotted in the (e, v)-plane for various values of the parameters μ and γ. Figure 4.13a, with $\mu = 0$ and regressive expectations ($\gamma > 0$), is unstable, with the system moving to the east, tending asymptotically to the constant-v locus, with ever-increasing expected rate of change of price, e. The "tangible" sector of the economy will tend to a situation where the resource market clears, with $R^D(v) = R_1^S$. Figure 5.13f is similarly unstable, with negative and ever-decreasing expected rate of change of price and the economy tending asymptotically to the constant-v locus. The tangible sector of the economy will tend to a situation where the resource market clears, with $R^D(v) = R_2^S$. For $\mu \neq 0$ we see that $e^* = \gamma r/\mu$ is finite: with regressive expectations e* is positive; with progressive expectations, negative.

In previous encounters with the quasi-Hotelling resource flow supply function, at quasi-equilibrium e* has been zero, and $R^{HS}(e^*) = R_2^S$, and $R_e^{HS}(e^*) = 0$. But with compound adaptive expectations, e* is not necessarily zero. When

$$(5.91)\qquad 0 < r - a < e^* < r + a,$$

the supply of resource flow at quasi-equilibrium is elastic, with

$$(5.92) \qquad R_e^{HS}(e^*) = -b < 0, \qquad -a < e^* - r < a.$$

This can occur only with regressive expectations (cases (b), (c), and (d) of Figure 5.13); with progressive expectations $e^* < 0$ and the supply of resource flow at quasi-equilibrium is inelastic.

The two-dimensional system can be written as

$$(5.93) \qquad \dot{e} = (\mu - \gamma)g - \mu e + \gamma r, \qquad \mu > 0,$$

$$g = \lambda_V(R^D(v) - R^{HS}(e)).$$

Analysis in Appendix D7 shows that in cases (b), (c), and (e) the system is definitely stable, in cases (a), (f), (g), (g) definitely unstable, and in case (d) stable iff

$$(5.94) \qquad (\mu - \gamma)b < \mu/\lambda_V - v^* R_v^D.$$

Allowing money wage W and money price of output P to vary in response to excess effective demands in the markets for labour and output respectively, leads to the three-dimensional system of equations (5.90). Analysis in Appendix D7 shows that for the quasi-equilibrium Q^{DC}, in the region of Keynesian unemployment, cases (b), (c), and (e) are definitely stable, and case (d) possibly unstable. For quasi-equilibrium Q^{NRC}, in the region of repressed inflation, analysis shows that cases (c) and (e) are definitely stable, and cases (b) and (d) possibly unstable. For both quasi-equilibria, cases (a), (f), (g), (h) are definitely unstable.

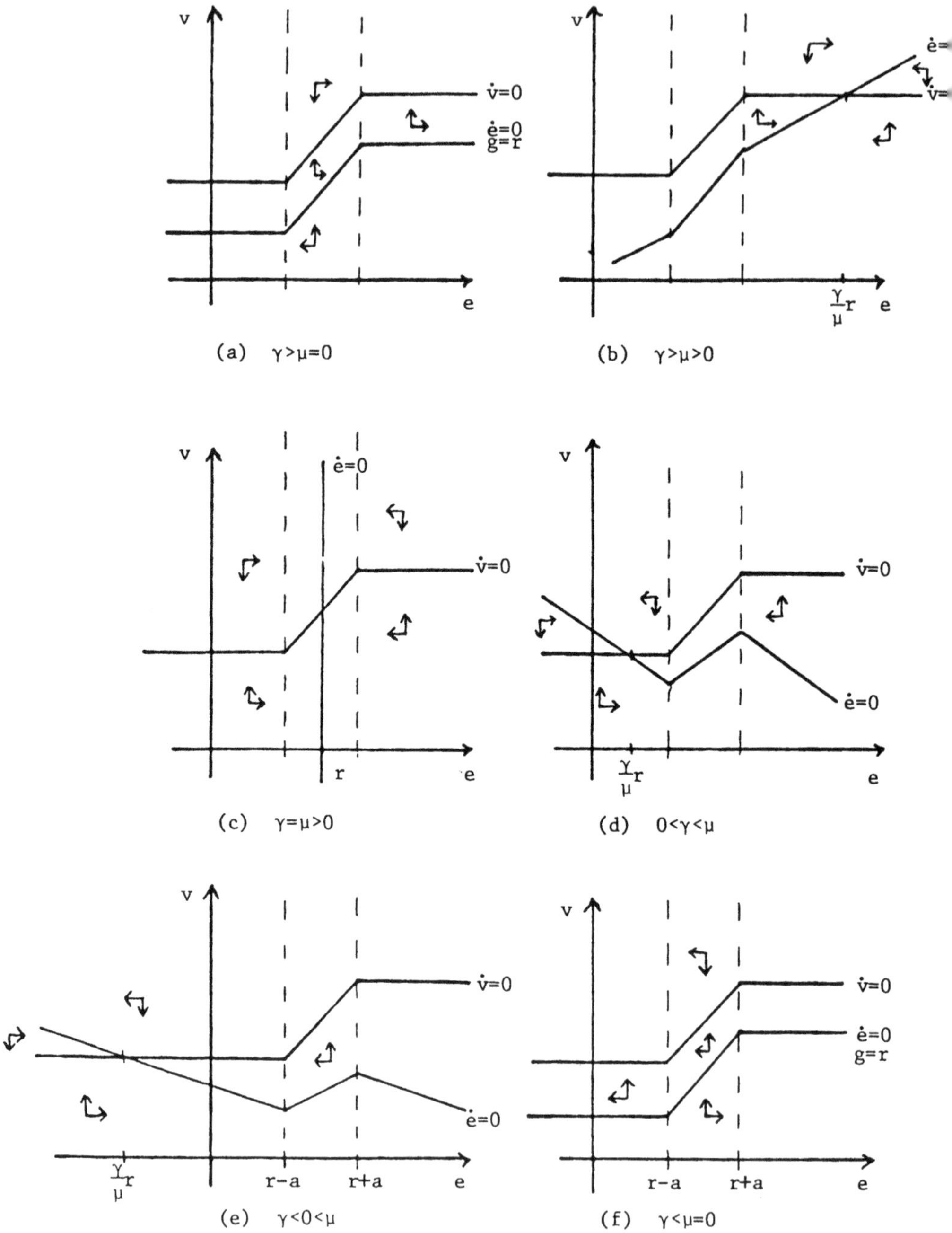

Figure 5.13: Phase diagrams, quasi-Hotelling resource supply, $g = \lambda_V R^X$, compound adaptive expecations.

5.8. An expectational formulation of price adjustment.

In Chapter IV we considered three possible formulations of price adjustment: the simple formulation of equations (4.7) used in this chapter, the alternative formulation of equations (4.18), and the Phelps-Friedman formulation of equations (4.14). In discussing the three, we gave motivations for each, rejecting the Phelps-Friedman because the alternative formulation seemed to describe the same behaviour more conveniently, and rejecting the alternative (using Occam's razor) after we had found that the simple formulation led to the same stability properties, and was simpler.

In this section we introduce a formulation which has often been postulated in models of optimal growth with non-renewable natural resources, in which expectations of the rate of change of real resource price directly affect the actual rate of change of real resource price:

$$(5.95) \qquad g = \lambda_V R^X - \lambda_P Y^X + e.$$

That is, even if the markets for resource flow and output are clearing (an assumption of, for instance, Stiglitz (1974b)), the real price will change at a rate equal to the expected rate. Then, if resource owners expect the Hotelling principle to hold, it can, even if all markets clear.

Although this formulation is usually presented without discussion or motivation, or sometimes not presented at all, but hidden implicitly in the model, we can argue some justification for it, other than existence of the Hotelling principle, from the micro conditions of the economy. The formulation of equation (5.95) would result from

$$(5.96) \qquad \begin{aligned} \dot{V}/V &= \lambda_V R^X + \dot{V}^e/V, \\ \dot{P}/P &= \lambda_P Y^X + p, \\ e &= \dot{V}^e/V - p, \\ p &\equiv \dot{P}^e/P. \end{aligned}$$

In equations (5.96), $\dot{V}^e/V$ is the expected rate of change of the money price of resource flow, p is the expected rate of change of the money price of output (a proxy for the price level). The second equation of (5.96) can be thought of as the outcome of the following process: in the market for output there are two direct forces on the price level--imbalance of the market for output and participants' expectations of future price inflation, which through their behaviour will affect the price of output, even if the output market clears. Similarly, in the first equation of (5.96) we can think of the direct effects of imbalance of the market for resource flow and participants' expectations of the future money price increases on the market. Subtracting the second from the first leads to the formulation of equation (5.95).

The motivation for the expectational price adjustment formulation of equation (5.95) is not as strong as for the simple formulation used previously in this chapter, but the expectational formulation (5.95) has been used in the literature to lead to economies which obey the Hotelling principle. We shall see that, in our model, with the compound adaptive expectations formulation of section 5.7, it still cannot save the day.

5.8.1. <u>The simple resource supply function</u>.

The case of the simple resource flow supply function, the compound adaptive expectations formulation, and the expectational formulation of price adjustment. The system can be described as

$$(5.97) \qquad \dot{e} = (\mu - \gamma)g - \mu e + \gamma r, \qquad \mu > 0,$$

$$g = \lambda_V R^X - \lambda_P Y^X + e,$$

$$h = \lambda_W N^X - \lambda_P Y^X,$$

$$R^S = R^{SS}(e).$$

This is a three-dimensional system, but for the moment we assume that P and W are unchanging. The resulting two-dimensional system can be written as

$$(5.98) \qquad \dot{e} = (\mu - \gamma)g - \mu e + \gamma r, \qquad \mu > 0,$$

$$g = \lambda_V (R^D(v) - R^{SS}(e)) + e.$$

We have not plotted this system explicitly, but Figure 5.12, for the system of equations (5.88) with the simple formulation of price adjustment, is similar to such a plot. Totally differentiating the second of the equations of (5.88) and (5.98), we obtain

$$(5.99) \qquad \left.\frac{dv}{de}\right|_{g}^{5.88} = \lambda_V R_e^{SS} / \lambda_V R_v^D > 0, \text{ and}$$

$$\left.\frac{dv}{de}\right|_{g}^{5.98} = (\lambda_V R_e^{SS} - 1)/\lambda_V R_v^D > \left.\frac{dv}{de}\right|_{g}^{5.88}.$$

Thus the constant-v locus, $\dot{v} = 0$, would be slightly steeper in a plot of the system of equations (5.98) than in a plot of the system of equations (5.88). Figure 5.12 will double for both.

As argued in section 5.7.1, Figure 5.12a, with $\mu = 0$ and regressive expectations ($\gamma > 0$), is unstable, with the economy moving to the NE, between the constant-v and constant-e loci, with $\dot{e}$ positive and less than γr, and g positive and less than r. Figure 5.12f is similarly unstable, except that expectations are progressive, with $\gamma < 0$. For $\mu \neq 0$ we see that $e^* = \gamma r/\mu$ is finite. With regressive expectations e^* is positive, with progressive expectations, negative.

Analysis in Appendix D8 shows that cases (a), (f), (g), (h) are unstable and that the system of equations (5.98) is locally dynamically stable iff inequality (5.100) holds (from inequality (D97))

$$(5.100) \qquad -(\mu - \gamma)\ \lambda_V\ R_e^{SS} - \gamma < -v^*\ \lambda_V\ R_v^D.$$

Cases (b) and (c) are shown to be definitely stable, cases (d) and (e) are possibly unstable.

Allowing money wage W and money price of output P to vary in response to excess effective demands in the markets for labour and output respectively leads to the three-dimensional system of equations (5.97). Analysis in Appendix D8, for the quasi-equilibrium Q^{DC}, in the region of Keynesian unemployment, shows that it is definitely stable in cases (b) and (c), definitely unstable in cases (a), (f), (g), (h), and possibly unstable in cases (d) and (e). The quasi-equilibrium Q^{NRC}, in the region of repressed inflation, is definitely stable for case (c), definitely unstable for cases (a), (f), (g), (h), and possibly unstable in cases (b), (d), and (e). These stability properties are identical with those of section 5.7.1, where there was no direct effect of expectations on price changes.

5.8.2. The quasi-Hotelling resource supply function.

The case of the quasi-Hotelling resource flow supply function, the compound adaptive expectations formulation, and the expectational formulation of price adjustment. The system can be described as

$$\text{(5.101)} \qquad \dot{e} = (\mu - \gamma)g - \mu e + \gamma r, \qquad \mu > 0,$$

$$g = \lambda_V R^X - \lambda_P Y^X + e,$$

$$h = \lambda_W N^X - \lambda_P Y^X,$$

$$R^S = R^{HS}(e).$$

This is a three-dimensional system, but for the moment we assume that P and W are unchanging. The resulting two-dimensional system can be written as

$$\text{(5.102)} \qquad \dot{e} = (\mu - \gamma)g - \mu e + \gamma r, \qquad \mu > 0,$$

$$g = \lambda_V(R^D(v) - R^{HS}(e)) + e.$$

Figure 5.14 shows this system plotted in the (e, v)-plane for various values of the parameters μ and γ. Comparing Figure 5.14 with Figure 5.13, we see that the constant-v locus (and hence, from the compound adaptive expectations formulation equation, the constant-e locus) is steeper in the former figure. This follows from equations (5.99) with $R^{HS}(e)$ replacing $R^{SS}(e)$. Figure 5.14a, with $\mu = 0$ and regressive expectations ($\gamma > 0$), is unstable, with the economy moving to the NE between the constant-v and constant-e loci, with $\dot{e}$ positive and less than r, and g positive and less than r, similar to the behaviour seen in Figure 5.12a. It is easily shown that if

$$(5.103) \qquad \lim_{v\to\infty}(v\, R_v^D) \equiv -c < 0$$

then

$$(5.104) \qquad \lim_{t\to\infty} g(t) \equiv \bar{g} = \gamma r/(\gamma + \lambda_V\, c) < r, \quad \text{and}$$

$$\lim_{t\to\infty} \dot{e}(t) \equiv \bar{\dot{e}} = \lambda_V\, c\, \bar{g} < \gamma r.$$

Figure 5.14f is similarly unstable, except that expectations are progressive and $\gamma < 0$. For $\mu \neq 0$ we see that $e^* = \gamma r/\mu$ is finite: with regressive expectations e* is positive, with progressive expectations, negative.

Analysis in Appendix D9 shows that cases (a), (f), (g), (h) are unstable and that the system of equations (5.102) is locally dynamically stable iff inequality (5.105) holds (from inequality (D117))

$$(5.105) \qquad -(\mu - \gamma)\, \lambda_V\, R_e^{HS}(e^*) - \gamma < -v^*\, \lambda_V\, R_v^D.$$

As discussed in section 5.7.2, when inequality (5.91) does not hold, R^{HS} is inelastic, but in this case (5.105) holds iff $\gamma > 0$, which includes case (a) and perhaps case (b). Case (b) can also result in inequality (5.105) being satisfied, and $R_e^{HS}(e^*)$ being negative, from equation (5.92). Analysis in Appendix D9 shows that cases (b) and (c) are definitely stable, and that cases (d) and (e) are stable iff inequality (5.105) holds. Cases (a), (f), (g), (h) are unstable.

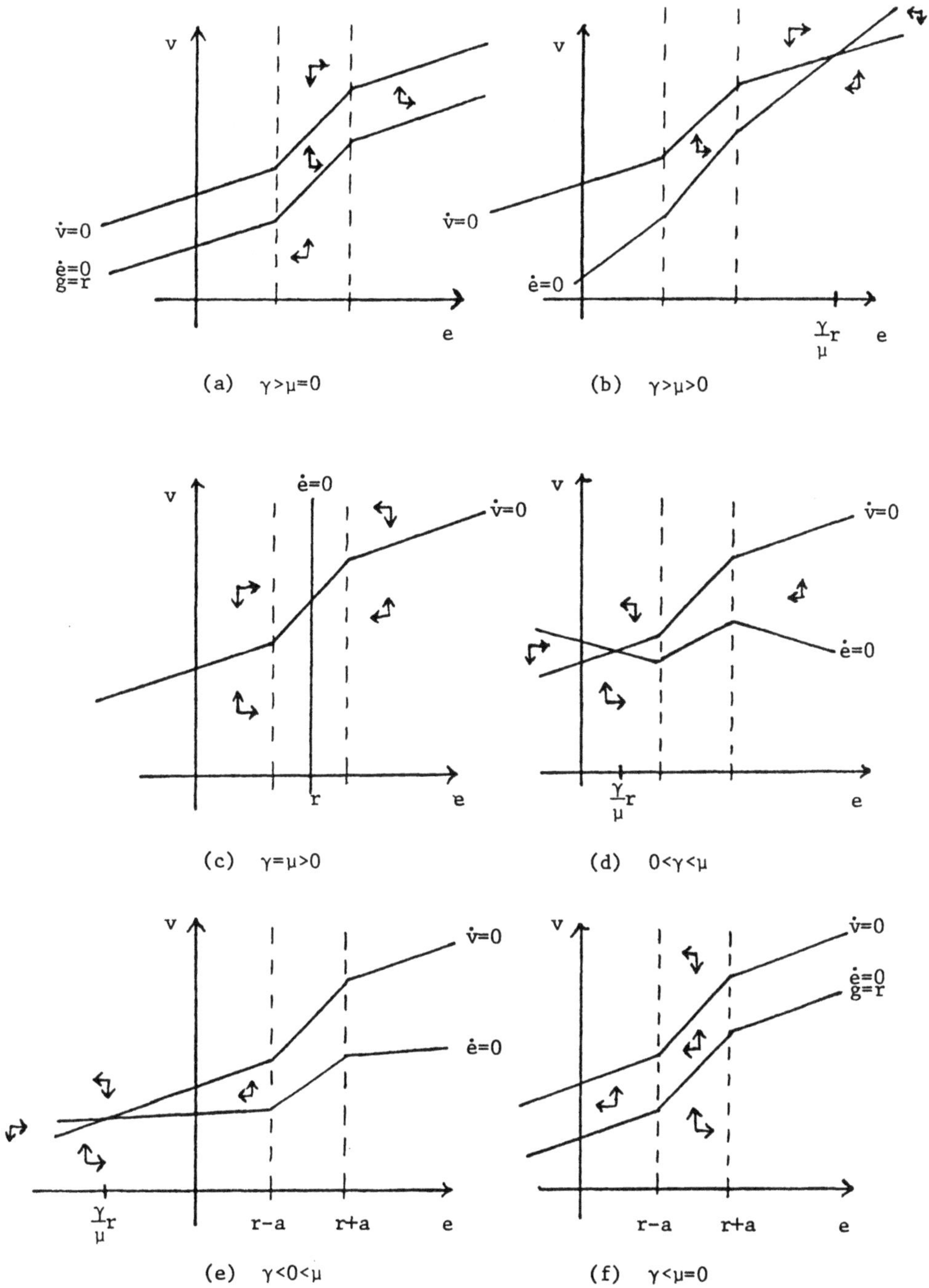

Figure 5.14: Phase diagrams, quasi-Hotelling supply, $g = \lambda_V R^X + e$, compound adaptive expectations.

Allowing money wage W and money price of output P to vary in response to excess effective demands in the markets for labour and output respectively leads to the three-dimensional system of equations (5.101). Analysis in Appendix D7 shows that for the quasi-equilibrium Q^{DC}, in the region of Keynesian unemployment, cases (b) and (c) are definitely stable, and cases (d) and (e) possibly unstable. For quasi-equilibrium Q^{NRC}, in the region of repressed inflation, analysis shows that case (c) is definitely stable, and cases (b), (d), and (e) possibly unstable. For both quasi-equilibria, cases (a), (f), (g), (h) are definitely unstable. These results are identical with those obtained in section 5.8.1, with the simple resource supply function, and with those obtained in section 5.7.1, with the simple resource supply function and the simple price adjustment formulation.

5.9. <u>Conclusion of the stability analysis.</u>

In sections 5.4 through 5.8 we have examined the stability of the three quasi-equilibria, Q^{DC}, Q^{DNR}, Q^{NRC}, which occur with the simple formulation of price adjustment as

$$(5.106) \qquad \bar{Y} \lesseqgtr F(\bar{N}^S, R^S(e^*)),$$

where (w*, v*, e*) are the real wage, real resource flow price, and expected rate of change of real resource flow price at quasi-equilibrium. We have examined the stability properties of Q^{DC}, in the region of Keynesian unemployment, and Q^{NRC}, in the region of repressed inflation, particularly closely, deriving sufficiency conditions for their local stability in the appendices. Because of the geometry of the market-clearing loci, Q^{DNR}, the point of triple market clearing, is a stable node: a branch of the constant-w locus is horizontal through the point, and a branch of the constant-v locus, vertical.

We have found conditions sufficient for the local stability of both quasi-equilibria under the eight possibilities of two resource supply functions and three formulations of adaptive expectations, and the final combination of compound adaptive expectations and the expectational formulation of price adjustment of section 5.8. Only when, in the compound adaptive expectations formulation, $\mu = 0$ did we find certain instability. With $\mu = 0$ the formulation becomes

$$(5.107) \qquad \dot{e} = \gamma(r - g), \qquad \gamma \gtrless 0.$$

In this equation the only negative feedback is through the effect of the expectations on the supply of resource flow and hence the imbalance of the resource flow market: with $\mu = 0$ there is no direct effect of e on $\dot{e}$. When the supply of resource flow is inelastic, there is no negative feedback, and the economy has no quasi-equilibrium, as depicted in cases (a) and (f) of Figures 5.13 and 5.14. Even when the supply of resource flow is elastic this can occur, as in cases (a) and (f) of Figure 5.12. This behaviour should not surprise us, but the formulation of equation (5.107) is not typical, since typically there will be a direct linkage, with participants comparing their expectations with their experience, and adjusting their expectations accordingly.

In response to the question of whether the system is less stable if resource owners have progressive expectations than if they have regressive expectations (that is, if they expect that deviations will be amplified rather than being attenuated), the analyses show that in the case of the price level expectation formulation of section 5.6, regressive expectations ($\alpha < 0$) are generally sufficient for stability of the quasi-equilibria. In section 5.7 with the first version of compound adaptive expectations (equation (5.85)), we find no general pattern: cases (b) and (d) are cases of regressive expectations ($\gamma > 0$), case (e) is a case of progressive expectations ($\gamma < 0$), cases (a) and (f) as discussed above result in the atypical formulation of equation (5.107) and both are always unstable for $\gamma \gtrless 0$, case (c) occurs when $\gamma = \mu$ and is regressive. Case (c) is always stable, but cases (b), (d), and (e) may be unstable, although we have derived sufficiency conditions for stable quasi-equilibria with each. With the second version of compound adaptive expectations (equation (5.86)), there is again no general pattern: cases (c) and (d) show regressive expectations ($\delta > 0$), cases (e), (f), (g), and (h) show progressive expectations ($\delta < 0$). Case (c) is always stable, cases (f), (g), and (h) are always unstable, and cases (d) and (e) may be unstable although we have derived sufficiency conditions for stable quasi-equilibria with each. In section 5.8 with the expectational formulation of price adjustment, again we find no general pattern: case (c) is always stable, cases (a) and (f) always unstable, and we have derived conditions sufficient for stability in the cases (b) and (d) of regressive expectations and case (e) of progressive expectations.

A final task is to consider the existence of equilibrium in the asset market (resulting in the Hotelling principle) and equilibrium in the three flow markets we have considered. We can speak of the asset market as being in equilibrium when the expected returns on all assets are equal (assuming either risk-neutral traders or that the expected return includes a risk premium). Whether or not this equilibrium is stable depends on the dynamic properties of the market: the expected return on each asset as a function of the holdings of that asset and all others. We have not been concerned with the whole asset market, merely the supply of and demand for holding stocks of non-renewable natural resources. We shall see that "equilibrium" in the asset market and the three flow markets cannot exist simultaneously in our model.

As we have noted, the only way a stock of non-renewable natural resources can produce a current return for its owner is by appreciating in value. Hence the anticipated rate of return on such stocks for risk-neutrality will be the expected proportional change of price e. We have assumed that the expected returns on other assets are equal and equal to the rate of interest, r, which has been taken as an exogenous parameter in the model. Other assets can include pure consumer loans: we do not have to assume varying capital equity for other assets to exist. If, as we have assumed, stocks of non-renewable natural resource are gross substitutes with other assets, then in equilibrium in the asset market

(5.108) $$e = r.$$

The long-run growth paths examined by such authors as Stiglitz (1974b) have assumed equilibrium in the asset market, which with fulfilled expectations or perfect myopic foresight leads to the condition that

(5.109) $$g = r.$$

What conditions are necessary for harmony between equilibrium in the asset market, and equilibrium in the other three markets?

Consider the equations of motion of the economy

$$(5.110) \qquad g = g(v, w, e), \qquad h = h(v, w, e).$$

Differentiating with respect to time, we obtain

$$(5.111) \qquad \begin{bmatrix} \dot{g} \\ \dot{h} \end{bmatrix} = \begin{bmatrix} g_v & g_w & g_e \\ h_v & h_w & h_e \end{bmatrix} \begin{bmatrix} \dot{v} \\ \dot{w} \\ \dot{e} \end{bmatrix} .$$

Let us require $\dot{e} = 0$ for equilibrium. From the asset market we have

$$(5.112) \qquad \dot{v}/v \equiv g = r, \quad \text{and} \quad \dot{g} = 0.$$

We can now ask what the conditions on J are for equation (5.112) to hold, with a similar form of exponential growth in real wage in the labour market:

$$(5.113) \qquad \dot{w}/w \equiv h = q, \quad q \text{ any real constant, and} \quad \dot{h} = 0,$$

where J is the Jacobian matrix of the partial derivatives of g and h with respect to v and w, first met in Appendix C.

Rewriting equation (5.111), with $\dot{e} = 0$ and equations (5.112) and (5.113), we are asking what the conditions on J are for

$$(5.114) \qquad \begin{bmatrix} 0 \\ 0 \end{bmatrix} = J \begin{bmatrix} vr \\ wq \end{bmatrix}$$

to hold. It is sufficient for J to be singular, for then

$$(5.115) \qquad J\,\underline{x} = \underline{0}, \qquad \text{for all } \underline{x}.$$

But, from equation (D22),

$$(5.116) \qquad |J| = C/w^*v^*\mu,$$

and as found in Appendix D2, C is positive in the case of both Q^{DC} and Q^{NRC}, iff μ is positive. Thus J is seen to be non-singular for both quasi-equilibria, and an equilibrium satisfying equations (5.112) and (5.113), that is, in which the Hotelling principle holds and the real wage is constant or changing exponentially (a reasonable definition of an equilibrium in the flow markets with asset market equilibrium), cannot exist.

Analysis in this chapter of the short-run equilibrium of this model has shown that equilibrium of the three flow markets cannot exist with equilibrium on the stock/asset market (implicit in the Hotelling principle). Stable equilibrium in the three flow markets has been shown to occur only with constant real resource flow price, although we saw that not all such equilibria are stable. Thus the analysis of this chapter shows that under reasonable conditions the long-run growth paths postulated by others, even if eventually unstable, cannot be supported by the micro-behaviour of the model in the short run. The Hotelling principle will not hold. This implies that it is not sufficient for efficient allocation of non-renewable natural resources by a competitive economy that participants have infinite perfect foresight or that a complete set of futures markets exists, since although any long-run path obeying the Hotelling principle would then be stable in the long-run sense, it would remain unstable in the short run.

CHAPTER VI: CONCLUSION

The objects of this study have been to examine how the existence of non-renewable natural resources affects the short-run behaviour of the economy, and to examine the possible micro-foundations for the Hotelling principle's holding in the short run. In order to do this we have built a short-run disequilibrium model where firms, households, workers, resource owners, and the government interact in the three flow markets explicitly examined (those for homogeneous output, labour, and resource flow), and the asset market which is assumed to exist, without being closely examined. We shall now briefly review the important characteristics of our model and the conclusions we have reached in relating the behaviour of the model to the real world. In doing so we shall mention how some of our assumptions could be relaxed to strengthen the model and lead to new directions for further work.

6.1. Discussion of assumptions and results.

In Chapter II we presented the basic model. Firms sell output to households, investment goods buyers, and the government on the output market. We make no distinction between households and workers: households sell labour services to firms on the labour market. Resource owners sell a flow of resource to firms on the market for resource flow. We assume that the net revenue (profit) of the firms accrues to the households, as does the return to resource owners. From their income, net of income taxes, households save and consume. Our treatment of this decision is weak, since rather than deriving their saving/consumption behaviour as the result of some optimizing decision on the part of the households, we use the standard assumption of constant and equal average and marginal propensities to save out of disposable income. This allows us to devote our energies to the behaviour of markets out of equilibrium and the effect of this on the participants and to the behaviour of the resource owners in forming expectations of the future and acting accordingly. We feel that the model is robust and that a fuller study of household consumption/savings behaviour would not result in great changes to the model, as neither would an elastic supply of labour. It would be interesting to see whether the neo-Keynesian assumption of different saving propensities among workers, capitalists, and resource suppliers would significantly alter the behaviour of the model.

The rate of return on the asset market is held constant throughout the study (perhaps through a permissive government monetary policy), as discussed below, and if the rate of return is an argument in the consumption function (perhaps leading to the Pigou or real balance effect), it is implicitly. Although the resource price varies in the model, it does not directly affect consumption. We have assumed exogenous demand for investment goods and exogenous government fiscal policy, although variations in these have been considered via changes in autonomous demand for output, a function of demand for investment goods, government expenditure, total income tax revenue, and the household propensity to save.

The driving force behind the basic model is the profit maximizing of the firms, which transform the resource flow and labour services to output, as described by the production function. Although we have been able to develop the model almost entirely without using a specific form of production function, we have assumed two characteristics which are basic to the study: the possibility of substitution between the two variable input factors of production, labour and resource, and diminishing returns to scale from the short-run assumption of fixed capital inputs. We have not closely analysed the implications of a fixed proportions production function, but it would lead to very different effective (and notional) market-clearing loci, since the complementary nature of fixed-proportions inputs means that if the representative firm were constrained on one input factor market (a sellers' market), the demand for the other input and the supply of output would be completely inelastic. Similarly, if the firm were constrained on the output market (a buyers' market), the demands for both factor inputs would be completely inelastic. With a constant-returns-to-scale production function, the SC region of classical unemployment in which the firm is unconstrained could not exist, since the level of activity of the economy would be indeterminate: the firm would buy more inputs and sell more outputs until one of the markets cleared and the firm became constrained. Clearly, if the region of classical unemployment is to be modelled, the production function cannot be constant-returns-to-scale, as is usually assumed in optimal growth theory (a linear, homogeneous production function).

Chapter II examines the adjustment of the tatonnement process, in which no non-market-clearing exchange takes place, or in which any contracts which are made before market clearing is achieved are recontractable. In a world of fixed contracts, this is obviously an unrealistic model, and Chapter III examines a model in which the existence of non-market-clearing trading results in income-constrained, effective schedules.

Previous studies have modelled one variable input (labour); ours is the first to model two, and hence the first to describe in detail those regions where the representative firm is constrained on one factor input market but not the other (RC, NC, DRC, and DNC). In particular we have found a region (DRC, with buyers' markets for output and labour as in Keynesian unemployment, but a sellers' market for resource flow) in which lowering the real wage would not reduce unemployment, and another (DC, the region of Keynesian unemployment), in which a fall in the real price of resource would result in increased unemployment as firms substituted (relatively) cheaper resource for dearer labour.

Chapter IV examines the adjustment of prices in response to unbalanced supply and demand. The assumption is made that quantitites adjust infinitely more rapidly than prices: we can thus ignore the adjustment of quantities, which are assumed to have reached their target values, functions of the two real prices, in a period of adjustment too short to concern us. This assumption can be justified in terms of fixed contracts, as has been done in previous studies of the labour market. We have made it not only for the labour market but also for the markets for output and resource flow. The relative behaviour of prices and quantities in these markets is amenable to empirical observation, and the hypothesis that quantities adjust "much" more rapidly than prices to empirical testing. It might also be possible to construct a model with simultaneous adjustment of prices and quantities to see whether slower quantity adjustment would affect our conclusions, if not in three markets, then perhaps in one or two at first.

Economic theorists are still developing hypotheses of the mechanism(s) of price adjustment to replace the essentially ad hoc Walrasian excess demand formulation. We have mentioned three, including the latter, and examined the consequences of two for the modelled economy as prices adjust. We have assumed that the price of resource flow determines the price of resource stocks, and that consequently the latter adjusts in response to imbalances in the flow market. In some models (of the capital market) stock/flow causation is in the opposite direction. The relationship between the two prices could bear closer examination.

Thus in the first three main chapters we have developed a model in which trading can take place in the three flow markets without market clearing, and have examined the quasi-equilibria which can occur if the supplies of the two factor input flows are inelastic, and if prices adjust at finite rates in response to imbalances of the markets. In particular we have examined the existence, uniqueness, and stability of these quasi-equilibria, and their comparative statics in response to changes in resource flow, autonomous demand, and to technical progress.

In Chapter V we have been concerned with making the model more realistic with respect to the supply of the flow of resource: we have modelled the resource owners as holding stocks of resource as one asset in a portfolio whose present value they attempt to maximize in response to their expectations of the future prices of resource. We formulate two resource flow supply functions, both non-increasing functions of the expected rate of change of resource price. We assume throughout that this expectation is held uniformly, although non-uniform expectations can be modelled implicitly in the resource flow supply function. In an area of economics which will benefit from further analysis and econometric study, we have shown that myopic perfect foresight is inadequate to describe the general behaviour we are examining and have developed two formulations of expectation formation, expectation of the price level (to model resource owners' ignorance of the Hotelling principle whereby the anticipated gain from holding stocks of non-renewable natural resource is equal to the return from holding other assets), and compound adaptive expectations (to model resource owners' awareness of the Hotelling principle). Both of these formulations can model progressive or regressive expectations. In the absence of extensive work by economists on the formation of expectations our formulations have been essentially ad hoc: after noting that such formulations might obey three "reasonable" basic properties, we have discussed the implications of each, and examined the stability of the possible quasi-equilibria in an attempt to determine whether the micro-foundations of any of our models will support an economy obeying the Hotelling principle in the short run.

Our answer is that the Hotelling principle, with the real price of resource growing at a proportional rate equal to the interest rate, cannot be supported by the short-run economy we have modelled, making various reasonable assumptions about the behaviour of the resource owners. The consequences of this for government policy will be discussed below.

Apart from any reservations the reader might have about the specific resource flow supply functions and expected resource price formulations we have used, the model is not completely a general equilibrium (or disequilibrium) model, since we treat the return to other assets, the interest rate r, as constant, and unaffected by fluctuations in the three flow markets we have modelled. This is unrealistic: the level of activity in the economy as modelled by the level of homogeneous output will definitely affect the rate of return on assets, particularly the rate of return on productive investments, as the marginal product of (fixed) capital varies. This is a shortcoming of the model, but inclusion of the link between level of aggregate output and the interest rate is likely to lead to further instability in the model, and so would not invalidate our conclusions.

In a final attempt to obtain an economy obeying the Hotelling principle, we have postulated a price adjustment formulation in which expectations of price change will lead directly to actual price change, even with general market clearing. This formulation results in stability properties of the quasi-equilibria virtually identical with those of the earlier formulations, thus even when autonomous price changes can occur (as a direct result of expectations of price changes) with general market clearing, the economy will not in general obey the Hotelling principle, which is incompatible with quasi-equilibrium.

One further shortcoming of the model is that it is deterministic. The resource supply functions have been able to embody uncertainty in the minds of the resource suppliers implicitly, but a thorough-going treatment of the model including uncertainty would be much different from our model, and might result in different conclusions. But in the absence of better hypotheses about expectation formation, and better understanding of the hold/sell decision of the resource supplier, formulation of the problem in probabilistic terms would result in effort beyond the point of diminishing returns. It has been suggested that one formulation of expectations should be the rational expectations hypothesis, but in a deterministic model rational expectations is identical with perfect foresight, which we have examined.

In response to the question of whether the system is more unstable if resource owners have progressive expectations (that is, if they expect any deviations to be amplified), in the case of the price level expectation formulation the answer is yes. In the case of compound adaptive expectations there is no definite answer.

6.2. The longer run.

To obtain a longer-run model we cannot simply speed up the processes of our model, since we have used an approach which is essentially that of the Hicksian period, most particularly in assuming that productive capital is fixed, and that the resource reserve-production ratio, if it affects price expectations, is constant in the short run.

The comparative statics analyses of Chapter IV give some idea of how the quasi-equilibria would move in the longer run in response to changes in variables we have taken to be exogenous and fixed. Discoveries of further stocks of resource would result in the resource flow supply curve's moving up: an increased flow of resource at every level of expected rate of change of price. With static expectations this translates into an increase in resource flow as examined in Chapter IV. An increase in fixed capital will be seen as a change in the production function. Our analyses of neutral and resource-augmenting technical change indicate how growth in the stock of capital would affect the model.

If resource price expectations are affected by changes in the resource reserve-production ratio, then the long-run behaviour of the model is not so easily seen. This effect could be modelled by adding a non-decreasing function of time onto the expectations formation equation, but in the short run this would be virtually constant. We have taken it as constant and zero, since in the period under consideration there would be a negligible change in the resource reserve-production ratio.

6.3. Policy recommendations.

Policy recommendations fall into two areas. In the theory of unemployment we have shown in Chapter III that there exist situations in which one of the factor input markets is a buyers' market and the other a sellers' market. In particular, if the markets for output and labour are buyers' markets as in the case of Keynesian unemployment, but the market for resource flow is a sellers' market (region DRC), then a reduction in the real wage would not reduce unemployment, contrary to traditional theory, since the demand for labour in this situation is inelastic with respect to the real wage, and any policy aimed at its reduction would be ineffective in lowering unemployment. But if government fiscal policy were altered to stimulate the level of autonomous demand for output, perhaps by increasing government expenditure and/or reducing taxes, the level of activity of the economy would be stimulated and with it the level of employment, as stated in conventional theory.

In the case of Keynesian unemployment, with all three markets buyers' markets, increased unemployment would result from a fall in the real price of resource, which would lead to substitution from (relatively) dearer labour to cheaper resource in the production process. Unemployment could be reduced by a fall in the real wage, as traditional theory predicts, and also by an increase in the level of autonomous demand for output.

In Chapter V we conclude that, in a short-run economy, we have been unable to find a reasonable micro-foundation for the Hotelling principle. We have noted that the principle is a necessary (although not sufficient) condition for the efficient intertemporal allocation of non-renewable natural resource. Having demonstrated, under reasonable conjectures of the price adjustment mechanism (including the direct effect of expectations), the resource flow supply function, and the mode of formation of expectations, that the Hotelling principle will not be obeyed in the short run, we conclude that with zero extraction costs a competitive market will not allocate non-renewable natural resource efficiently through time, at least not in the short run: if the price is too low there will be premature depletion, if too high, oversaving of resources.

It may be that unexamined factors will lead to a competitive economy which obeys the principle in the long run--the decreasing reserve-production ratio, perhaps. It may be that for particular resources, the assumption of a competitive market is unrealistic, and that monopolistic or oligopolistic markets will more closely obey the Hotelling principle. (This can be seen in the world market for oil, where the Organization of Petroleum Exporting Countries cartel has shifted the price of oil above its previous level, a level which in the U.S.A. in real terms had remained unchanged or even fallen in the previous twenty years.) But in the absence of such factors, this study provides a strong rationale for government action to control the rate of depletion of non-renewable natural resources. The whole study has been pitched in terms of price incentives for the participants in the economy, and one method of government action is intervention in the resource market to raise the price, perhaps by an excise tax on the sale of resource to encourage conservation. However, the same result could be achieved by non-price action, such as resource rationing, perhaps at some cost in terms of momentary efficiency of the economy.

BIBLIOGRAPHY

Alchian, A. A. (1970), "Information costs, pricing, and resource unemployment," in E. S. Phelps et al., Microeconomic Foundations of Employment and Inflation Theory, New York: Norton.

Arrow, K. J. (1959), "Towards a theory of price adjustment," in M. Abramovitz (ed.), The Allocation of Resources, Stanford: Stanford University Press.

Barro, R. J., and H. I. Grossman (1971), "A general disequilibrium model of income and employment," American Economic Review 61: 82-93.

Barro, R. J., and H. I. Grossman (1976), Money, Employment, and Inflation, New York: Cambridge University Press.

Benassy, J.-P. (1973), "Disequilibrium Theory," Ph.D. thesis, University of California, Berkeley.

Birkhoff, G., and S. MacLane (1977), A Survey of Modern Algebra, 4th ed., New York: Macmillan.

Clower, R. W. (1965), "The Keynesian counterrevolution," in F. Hahn and F. Brechling (eds.), The Theory of Interest Rates, London: Macmillan.

Cummings, R. G. (1969), "Some extensions of the economic theory of exhaustible resources," Western Economic Journal 7: 201-210.

Dasgupta, P., and G. M. Heal (1974), "The optimal depletion of exhaustible resources," Review of Economic Studies Symposium: 3-28.

Dasgupta, P., and J. E. Stiglitz (1975), "Uncertainty and the rate of extraction under alternative institutional arrangements," Technical Report No. 179, Institute of Mathematical Studies in the Social Sciences, Stanford University.

Drèze, J. M. (1975), "Existence of an exchange equilibrium under price rigidities and quantity rationing," International Economic Review 16: 301-320.

Friedman, M. (1968), "The role of monetary policy," American Economic Review 58: 1-17.

Garg, P. C. (1974), "Optimal economic growth with exhaustible resources," Ph.D. thesis, Stanford University.

Garg, P. C., and J. L. Sweeney (1978), "Optimal growth with depletable resources," Resources and Energy 1: 43-56.

Gilbert, R. (1976), "Optimal depletion of an uncertain stock," Technical Report No. 207, Institute for Mathematical Studies in the Social Sciences, Stanford University.

Gordon, R. L. (1967), "A reinterpretation of the pure theory of exhaustion," Journal of Political Economy 75: 274-286.

Grossman, H. I. (1971), "Money, interest, and prices in market disequilibrium," Journal of Political Economy 79: 943-961.

Grossman, H. I. (1972), "Was Keynes a 'Keynesian'?" Journal of Economic Literature 10: 26-30.

Hahn, F. H. (1966), "Equilibrium dynamics with heterogeneous capital goods," Quarterly Journal of Economics 80: 633-646.

Hahn, F. H. (1976a), "On non-Walrasian equilibria," Technical Report No. 203, Institute for Mathematical Studies in the Social Sciences, Stanford University.

Hahn, F. H. (1976b), "Keynesian economics and general equilibrium theory: reflections on some current debates," Technical Report No. 219, Institute for Mathematical Studies in the Social Sciences, Stanford University.

Hansen, Bent (1951), The Theory of Inflation, London: Allen & Unwin.

Hansen, Bent (1970), A Survey of General Equilibrium Systems, New York: McGraw-Hill.

Hanson, D. A. (1976), "Second best pricing policies for an exhaustible resource," presented at the AEA meetings, Atlantic City.

Heal, G. (1975), "Economic aspects of natural resource depletion," in D. W. Pearce and J. Rose (eds.), The Economics of Natural Resource Depletion, London: Macmillan.

Heal, G. (1976), "The relationship between price and extraction cost for a resource with a backstop technology," Bell Journal of Economics 7: 371-378.

Herfindahl, O. C. (1967), "Depletion and economic theory," in M. Gaffney (ed.), Extractive Resources and Taxation, Madison: University of Wisconsin Press.

Hicks, J. R. (1946), Value and Capital, 2nd ed., Oxford: Oxford University Press.

Hicks, J. R. (1965), Capital and Growth, Oxford: Oxford University Press.

Hotelling, H. (1931), "The economics of exhaustible resources," Journal of Political Economy 39: 137-175.

Iwai, K. (1974), "The firm in uncertain markets and its price, wage, and employment adjustment," Review of Economic Studies 41: 257-276.

Kaldor, N. (1939/40), "Speculation and economic stability," Review of Economic Studies 7: 1-27.

Kay, J. A., and J. A. Mirrlees (1975), "The desirability of natural resource depletion," in D. W. Pearce and J. Rose (eds.), The Economics of Natural Resource Depletion, London: Macmillan.

Koopmans, T. C. (1957), Three Essays on the State of Economic Science, New York: McGraw-Hill.

Koopmans, T. C. (1973), "Some observations on 'optimal' economic growth and exhaustible resources," in H. C. Bos et al. (eds.), Economic Structure and Development, Amsterdam: North-Holland.

Korliras, P. G. (1973), "Essays in disequilibrium macrodynamics," Ph.D. thesis, University of Rochester.

Leijonhufvud, A. (1968), On Keynesian Economics and the Economics of Keynes, New York: Oxford University Press.

Levhari, D., and N. Liviatan (1977), "Notes on Hotelling's economics of exhaustible resources," Canadian Journal of Economics 10: 177-192.

Loury, G. C. (1976), "The optimum exploitation of an unknown reserve," Discussion Paper No. 255, Center for Mathematical Studies in Economics and Management Science, Northwestern University.

Malinvaud, E. (1977), The Theory of Unemployment Reconsidered, Oxford: Basil Blackwell.

Negishi, T. (1962), "The stability of a competitive economy: a survey article," Econometrica 30: 635-669.

Nordhaus, W. D. (1973), "The allocation of energy resources," Brookings Papers in Economic Activity 3: 529-576.

Okun, A. (1975), "Inflation: its mechanics and welfare costs," Brookings Papers on Economic Activity 2: 351-402.

Patinkin, D. (1965), Money, Interest, and Prices, 2nd ed., New York: Harper & Row.

Peterson, F. M., and A. C. Fisher (1977), "The exploitation of extractive resources: a survey," Economic Journal 87: 681-721.

Phelps, E. S. (1968), "Money wage dynamics and labour market equilibrium," Journal of Political Economy 76: 678-711.

Robinson, J. (1977), "What are the questions?" Journal of Economic Literature 13: 1318-1373.

Shell, K., and J. E. Stiglitz (1967), "The allocation of investment in a dynamic economy," Quarterly Journal of Economics 81: 592-607.

Solow, R. M. (1974a), "The economics of resources or the resources of economics," American Economic Review 64: 1-14.

Solow, R. M. (1974b), "Intergenerational equity and exhaustible resources," Review of Economic Studies Symposium: 29-45.

Solow, R. M., and J. E. Stiglitz (1968), "Output, employment, and wages in the short run," Quarterly Journal of Economics 82: 537-560.

Stiglitz, J. E. (1974a), "Growth with exhaustible natural resources: efficient and optimal growth paths," Review of Economic Studies Symposium: 123-138.

Stiglitz, J. E. (1974b), "Growth with exhaustible natural resources: the competitive economy," Review of Economic Studies Symposium: 139-152.

Sweeney, J. L. (1977), "Economics of depletable resources: market forces and intertemporal bias," Review of Economic Studies 44: 125-141.

Van Order, R. (1976), "Excess demand and market adjustment," Economic Inquiry 14: 587-603.

Varian, H. (1976), "On balanced inflation," Economic Inquiry 14: 45-51.

Weinstein, M., and R. J. Zeckhauser (1975), "The optimal consumption of depletable natural resources," Quarterly Journal of Economics 89: 371-392.

APPENDIX A: APPENDIX TO CHAPTER II

A1: Appendix for section 2.2.1

The firm will maximize its profit

(A1) $$\pi = Y^S - wN^D - vR^D$$

subject to the constraint

(A2) $$Y^S = F(N^D, R^D)$$

where the production function $F(\cdot,\cdot)$ exhibits a positive and diminishing marginal product with respect to each input, and diminishing returns to scale. Labour services and resource flow are technical complements in production. The partial derivatives have the properties

(A3) $$F_N > 0, \quad F_R > 0, \quad F_{NN} < 0, \quad F_{RR} < 0, \quad F_{NR} > 0,$$

and we further assume that

(A4) $$F_{NR} < -F_{NN} \quad \text{and} \quad F_{NR} < -F_{RR},$$

so that

(A5) $$D_1 \equiv F_{NN}\, F_{RR} - (F_{NR})^2 > 0.$$

The production function is then strictly concave and the first-order conditions are necessary and sufficient for profit maximization:

(A6) $$w = F_N(N^D, R^D)$$

$$v = F_R(N^D, R^D).$$

Hence

$$\text{(A7)} \qquad N^D = N^D(w, v)$$

$$R^D = R^D(w, v)$$

$$Y^S = F(N^D, R^D) = Y^S(w, v).$$

To obtain the partial derivatives of these derived functions, differentiate the first-order conditions totally with respect to w and v. Hence, in matrix form

$$\text{(A8)} \qquad \begin{bmatrix} F_{NN} & F_{NR} \\ F_{NR} & F_{RR} \end{bmatrix} \begin{bmatrix} N_w^D & N_v^D \\ R_w^D & R_v^D \end{bmatrix} = I$$

$$\therefore \quad \begin{bmatrix} N_w^D & N_v^D \\ R_w^D & R_v^D \end{bmatrix} = \frac{1}{D_1} \begin{bmatrix} F_{RR} & -F_{NR} \\ -F_{NR} & F_{NN} \end{bmatrix}$$

which leads to the partial derivatives (equation (2.12)) in section 2.2.1. Further, we can consider

$$Y^S = F(N^D, R^D)$$

so that

$$\begin{bmatrix} Y_w^S \\ Y_v^S \end{bmatrix} = [F_N \quad F_R] \begin{bmatrix} N_w^D & N_v^D \\ R_w^D & R_v^D \end{bmatrix}$$

$$\text{(A9)} \qquad \therefore \quad Y_w^S = (w\, F_{RR} - v\, F_{NR})/D_1 < 0$$

$$Y_v^S = (-wF_{NR} + v\, F_{NN})/D_1 < 0$$

as stated in section 2.2.1.

A2: Appendix to section 2.3.1.

Consider the conditions for clearing in each of the three markets

(A10) $$N^D(w, v) - \bar{N}^S = 0,$$

$$R^D(w, v) - \bar{R}^S = 0,$$

$$Y^D - Y^S(w, v) = s(\bar{Y} - Y^S(w, v)) = 0,$$

where the autonomous demand for output $\bar{Y}$ is defined (equation 2.27)

(A11) $$\bar{Y} \equiv (I + G - (1 - s)T)/s.$$

Total differentiation and equations (A8) and (A9) yield

(A12) $$\left.\frac{dw}{dv}\right|_{N^X} = F_{NR}/F_{RR} < 0$$

$$\left.\frac{dw}{dv}\right|_{R^X} = F_{NN}/F_{NR} < 0$$

$$\left.\frac{dw}{dv}\right|_{Y^X} = -(vF_{NN} - wF_{NR})/(wF_{RR} - vF_{NR}) < 0.$$

From these equations, if inequalities (A4) hold

(A13) $$0 < F_{NR} < |F_{NN}|, |F_{RR}|,$$

then the labour-market-clearing locus in Figure 2.5 will be shallowly negatively sloped and the resource-market-clearing locus will be steeply negatively sloped. Inequalities (A13) state that the marginal product of each factor input diminishes more rapidly as the level of that factor increases than it increases with an increase in the other factor input, ceteris paribus.

With the Cobb-Douglas production function

$$\text{(A14)} \qquad Y = F(N, R) = N^{\beta}R^{\gamma} \qquad \beta+\gamma=1-\alpha;\ \alpha,\beta,\gamma > 0,$$

the slopes can be written

$$\text{(A15)} \qquad \left.\frac{dw}{dv}\right|_{N^X} = -\beta R/(\alpha+\beta)N,$$

$$\left.\frac{dw}{dv}\right|_{R^X} = -(\alpha+\gamma)R/\gamma N,$$

$$\left.\frac{dw}{dv}\right|_{Y^X} = -R/N,$$

whence we see that

$$\text{(A16)} \qquad \left.\frac{dw}{dv}\right|_{R^X} < \left.\frac{dw}{dv}\right|_{Y^X} < \left.\frac{dw}{dv}\right|_{N^X} < 0.$$

These inequalities hold for any possible R and N and so we see that there is only point of intersection and the relative slopes are as shown in Figure 2.5.

A3: Appendix for section 2.3.2

Consider the market-clearing conditions

$$\text{(A17)} \qquad \bar{N}^S - N^D(w^*, v^*) = 0$$

$$\bar{R}^S - R^D(w^*, v^*) = 0.$$

Differentiating totally with respect to all variables, we obtain

$$\text{(A18)} \qquad \begin{bmatrix} N_w^D & N_v^D \\ R_w^D & R_v^D \end{bmatrix} \begin{bmatrix} dw^* \\ dv^* \end{bmatrix} = \begin{bmatrix} d\bar{N}^S - dN^D \\ d\bar{R}^S - dR^D \end{bmatrix}$$

$$\therefore \quad \begin{bmatrix} dw^* \\ dv^* \end{bmatrix} = \begin{bmatrix} R_v^D & -N_v^D \\ -R_w^D & N_w^D \end{bmatrix} \begin{bmatrix} d\bar{N}^S - dN^D \\ d\bar{R}^S - dR^D \end{bmatrix} D_1.$$

These equations tell us how the market-clearing real prices w* and v* will change in response to either a change in the exogenous supplies, $\bar{N}^S$ or $\bar{R}^S$, or a change in the derived demand curves, N^D or R^D.

Thus, in the case of $d\bar{R}^S > 0$, and $dR^D = dN^D = d\bar{N}^S = 0$, we see that

$$\text{(A19)} \qquad dw^* = -N_v^D \, d\bar{R}^S \, D_1 > 0, \text{ and}$$

$$dv^* = N_w^D \, d\bar{R}^S \, D_1 < 0,$$

as argued in section 2.3.2.

Technical change will shift the derived demand curves, N^D and R^D, but how? We show below that for neutral or resource-augmenting technical change, dN^D will always be positive: there is a greater demand for labour at any (w, v), and that except for large α with resource-augmenting technical change (i), dR^D will always be positive. We then consider the signs of the changes in w* and v* due to technical change.

Consider the production function for neutral technical change

$$\text{(A20)} \qquad F^2(N, R) \equiv \alpha F(N, R), \qquad \alpha > 1.$$

The first-order conditions can be written

$$\text{(A21)} \qquad w = \alpha F_N(N^D, R^D)$$

$$v = \alpha F_R(N^D, R^D).$$

Differentiating totally with respect to α, N^D, and R^D, we obtain

$$\text{(A22)} \qquad \begin{bmatrix} F_{NN} & F_{NR} \\ F_{NR} & F_{RR} \end{bmatrix} \begin{bmatrix} dN^D \\ dR^D \end{bmatrix} = -\frac{d\alpha}{\alpha} \begin{bmatrix} F_N \\ F_R \end{bmatrix}$$

$$\therefore \begin{bmatrix} dN^D \\ dR^D \end{bmatrix} = -d\alpha \frac{1}{\alpha^2 D_1} \begin{bmatrix} F_{RR} & -F_{NR} \\ -F_{NR} & F_{NN} \end{bmatrix} \begin{bmatrix} w \\ v \end{bmatrix},$$

from which it follows that $dN^D > 0$ and $dR^D > 0$. From above

$$\text{(A23)} \qquad dw^* = -R_v^D \, dN^D + N_v^D \, dR^D$$

$$= w \, d\alpha/\alpha^3 \, D_1 > 0$$

and

$$\text{(A24)} \qquad dv^* = R_w^D \, dN^D - N_w^D \, dR^D$$

$$= v \, d\alpha/\alpha^3 \, D_1 > 0.$$

Thus in the case (ii) of neutral technical progress both market-clearing real prices increase as seen in Figure 2.8.

Consider the production function for resource-augmenting technical change

$$(A25) \qquad F^1(N, R) \equiv F(N, \alpha R), \qquad \alpha > 1.$$

The first-order conditions can be written

$$(A26) \qquad w = F_N(N^D, \alpha R^D)$$

$$v = \alpha F_R(N^D, \alpha R^D).$$

Differentiating totally with respect to α, N^D, and R^D, we obtain

$$(A27) \qquad \begin{bmatrix} F_{NN} & \alpha F_{NR} \\ \alpha F_{NR} & \alpha^2 F_{RR} \end{bmatrix} \begin{bmatrix} dN^D \\ dR^D \end{bmatrix} = -d\alpha \begin{bmatrix} F_{NR} R^D \\ \alpha F_{RR} R^D + F_R \end{bmatrix}$$

$$\therefore \begin{bmatrix} dN^D \\ dR^D \end{bmatrix} = -d\alpha \frac{1}{\alpha^2 D_1} \begin{bmatrix} \alpha^2 F_{RR} & -\alpha F_{NR} \\ -\alpha F_{NR} & F_{NN} \end{bmatrix} \begin{bmatrix} F_{NR} R^D \\ \alpha F_{RR} R^D + F_R \end{bmatrix}$$

$$(A28) \qquad \therefore \quad dN^D = v\, F_{NR}\, d\alpha/\alpha^2 D_1 > 0, \quad \text{and}$$

$$dR^D = -(\alpha R^D D_1 + F_{NN} F_R) d\alpha/\alpha^2 D_1 \gtrless 0.$$

Substituting into the analysis above

$$(A29) \qquad dw^* = -R_v^D\, dN^D + N_v^D\, dR^D$$

$$= F_{NR} R^D\, d\alpha/\alpha^2 D_1 > 0$$

and

(A28) $$dv^* = R_w^D \, dN^D - N_w^D \, dR^D$$

$$= (F_R + \alpha F_{RR} R^D) \, d\alpha/\alpha^2 D_1$$

$$> 0 \text{ if } 1/\alpha > - F_{RR} R^D/F_R$$

so that the real wage for market clearing always increases, and the real resource flow price for market clearing increases if $1/\alpha$ is greater than the resource elasticity of the marginal product of resource.

A4: Appendix for section 2.4.3

The dynamic system can be described by the following equations:

(A29) $$h \equiv \dot{w}/w = \lambda_W N^X - \lambda_P Y^X$$

$$g \equiv \dot{v}/v = \lambda_V R^X - \lambda_P Y^X$$

where the excess demands are given by

(A30) $$N^X = N^D - \bar{N}^S$$

$$R^X = R^D - \bar{R}^S$$

$$Y^X = Y^D - Y^S = s(\bar{Y} - Y^S)$$

The characteristic equation of the system is

(A31) $$\begin{vmatrix} w^*h_w - Z & w^*h_v \\ v^*g_w & v^*g_v - Z \end{vmatrix} = 0$$

and the system exhibits local dynamic stability in the region of general market clearing (w^*, v^*) iff the real parts of the roots of the characteristic equation are negative, $Re(Z) < 0$. It is necessary and sufficient for this that the trace of the Jacobian matrix of the partial derivatives of $\dot{w}$ and $\dot{v}$ is negative and that its determinant is positive. That is, that $B > 0$ and $C > 0$, where

$$\text{(A34)} \qquad B = -(w^*h_w + v^*g_v)$$

$$C = w^*v^*(h_wg_v - h_vg_w).$$

We see that

$$\text{(A35)} \qquad B = -(w^*(\lambda_W N_w^D + \lambda_P \; s \; Y_w^S) + v^*(\lambda_V R_v^D + \lambda_P \; s \; Y_v^S)) > 0$$

and

$$\text{(A36)} \qquad C/w^*v^* = \lambda_W\lambda_V(N_w^D R_v^D - N_v^D R_w^D)$$

$$+ \; s \; \lambda_W\lambda_P(N_w^D Y_v^S - N_v^D Y_w^S)$$

$$+ \; s \; \lambda_V\lambda_P(R_v^D Y_w^S - R_w^D Y_v^S)$$

$$= D_1(\lambda_W\lambda_V - vs \; \lambda_W\lambda_P - ws \; \lambda_V\lambda_P)$$

so that $C > 0$ if s is small or if λ_P much less than λ_W or λ_V.

APPENDIX B: APPENDIX TO CHAPTER III

B1: Appendix to section 3.6.

The eight situations described above can be derived from the solution of a constrained maximization problem as faced by the representative firm:

(B1) $$\max \pi = Y - wN - vR, \text{ such that}$$

$$Y = F(N, R)$$

$$N \leq \bar{N}^S$$

$$R \leq \bar{R}^S$$

$$Y \leq \bar{Y}.$$

Solution of this will automatically yield actual levels of quantitites transacted which obey the voluntary exchange assumption, although further analysis will be necessary to determine the effective demand (supply) on the "long" side of the market.

The Lagrangian is:

(B2) $$L = Y - wN - vR - \lambda_1(Y - F(N, R)) - \lambda_2(N - \bar{N}^S) - \lambda_3(R - \bar{R}^S) - \lambda_4(Y - \bar{Y}).$$

Differentiation of L with respect to Y, N, and R gives the Kuhn-Tucker conditions:

(B3) $$\lambda_1 + \lambda_4 = 1, \qquad \lambda_1 \geq 0$$

$$\lambda_1 F_N = w + \lambda_2$$

$$\text{(B3)} \qquad \lambda_1 F_R = v + \lambda_3$$

$$Y = F(N, R)$$

$$\lambda_2(N - \bar{N}^S) = 0, \qquad \lambda_2 \geq 0$$

$$\lambda_3(R - \bar{R}^S) = 0, \qquad \lambda_3 \geq 0$$

$$\lambda_4(Y - \bar{Y}) = 0, \qquad \lambda_4 \geq 0.$$

Then the eight situations can be characterized by the possibilities of the Kuhn-Tucker conditions.

(i) With $\lambda_2 = \lambda_3 = \lambda_4 = 0$, the firm is unconstrainted on any market. This is case SC. Solution of the conditions yields

$$\text{(B4)} \qquad N = N^{SCD} < \bar{N}^S$$

$$R = R^{SCD} < \bar{R}^S$$

$$Y = F(N, R) = Y^{SCS} < \bar{Y}$$

where

$$\text{(B5)} \qquad F_N(N, R) = w$$

$$F_R(N, R) = v$$

which are solved to obtain N and R as in the basic model in Chapter II

$$\text{(B6)} \qquad N = N^{SCD}(w, v)$$

$$R = R^{SCD}(w, v)$$

$$Y = F(N, R) = Y^{SCS}(w, v)$$

with the comparative statics equations derived in Appendix A1.

(B7)

$$N_w^{SCD} = F_{RR}/D_1 < 0$$

$$N_v^{SCD} = -F_{NR}/D_1 < 0$$

$$R_w^{SCD} = -F_{NR}/D_1 < 0$$

$$R_v^{SCD} = F_{NN}/D_1 < 0$$

$$Y_w^{SCS} = (wF_{RR} - vF_{NR})/D_1 < 0$$

$$Y_v^{SCS} = (vF_{NN} - wF_{NR})/D_1 < 0$$

$$D_1 = F_{NN}F_{RR} - (F_{NR})^2 > 0.$$

(ii) With $\lambda_4 = \lambda_2 = 0$, $\lambda_3 > 0$, the firm is constrained on the resource market, but unconstrained on the other two. This is the case RC. Solution of the conditions yields

(B8)

$$N = N^{RCD} < \bar{N}^S$$

$$R = \bar{R}^S < R^{SCD}$$

$$Y = F(N, R) = Y^{RCS} < \bar{Y}$$

with

(B9)

$$F_N(N, R) = w$$

$$F_R(N, R) > v$$

which can be solved to obtain N in terms of $\bar{R}^S$ and w

(B10)

$$N = N^{RCD}(w, \bar{R}^S)$$

(B10) $$R = \bar{R}^S$$

$$Y = F(N, R) = Y^{RCS}(w, \bar{R}^S).$$

Standard analysis yields the comparative statics equations:

(B11) $$N_w^{RCD} = 1/F_{NN} < 0$$

$$N_v^{RCD} = 0$$

$$Y_w^{RCD} = F_N/F_{NN} < 0$$

$$Y_v^{RCD} = 0.$$

(iii) With $\lambda_4 = \lambda_3 = 0$, $\lambda_2 > 0$, the firm is constrained on the labour market, but unconstrained on the other two. This is the case NC. Solution of the conditions yields

(B12) $$N = \bar{N}^S < N^{SCD}$$

$$R = R^{NCD} < \bar{R}^S$$

$$Y = F(N, R) = Y^{NCS} < \bar{Y}$$

with

(B13) $$F_N(N, R) > w$$

$$F_R(N, R) = v$$

which can be solved to obtain R in terms of $\bar{N}^S$ and v

(B14) $$N = \bar{N}^S$$

$$R = R^{NCD}(v, \bar{N}^S)$$

$$Y = F(N, R) = Y^{NCS}(v, \bar{N}^S).$$

Standard analysis yields the comparative statics equations

(B15) $$R_w^{NCD} = 0$$

$$R_v^{NCD} = 1/F_{RR} < 0$$

$$Y_w^{NCS} = 0$$

$$Y_v^{NCS} = F_R/F_{RR} < 0.$$

(iv) With $\lambda_2 > 0$, $\lambda_3 > 0$, $\lambda_4 = 0$, the firm is constrained on both factor input markets, but not on the output market. This is case NRC. Solution of the conditions yields

(B16) $$N = \bar{N}^S < N^{RCD}$$

$$R = \bar{R}^S < R^{NCD}$$

$$Y = F(N, R) = Y^{NRCS}(\bar{N}^S, \bar{R}^S) < \bar{Y}$$

with

(B17) $$F_N(N, R) > w$$

$$F_R(N, R) > v.$$

Standard analysis yields the comparative statics equations

(B18) $$Y_w^{NRCS} = 0$$

$$Y_v^{NRCS} = 0.$$

This case can only occur when

(B19) $$\bar{Y} > F(\bar{N}^S, \bar{R}^S).$$

(v) With $\lambda_2 = \lambda_3 = 0$, $\lambda_4 > 0$, the firm is constrained on the output market, but unconstrained on the other two. This is the case DC.

Solution of the conditions yields

$$\text{(B20)} \qquad N = N^{DCD} < \bar{N}^{S}$$

$$R = R^{DCD} < \bar{R}^{S}$$

$$Y = F(N, R) = \bar{Y} < Y^{SCS}$$

with

$$\text{(B21)} \qquad F_N(N, R) = w/\lambda_1$$

$$F_R(N, R) = v/\lambda_1, \qquad 0 < \lambda_1 \leq 1,$$

which can be solved to obtain N and R in terms of w/v and $\bar{Y}$

$$\text{(B22)} \qquad N = N^{DCD} < \bar{N}^{S}$$

$$R = R^{DCD}(w/v, \bar{Y})$$

$$Y = \bar{Y}.$$

Second-order conditions for profit maximization in the DC case are that the bordered Hessian determinant be positive

$$\text{(B23)} \qquad \begin{vmatrix} F_{NN} & F_{NR} & -w \\ F_{NR} & F_{RR} & -v \\ -w & -v & 0 \end{vmatrix} = 2wv\, F_{NR} - w^2\, F_{RR} - v^2\, F_{NN} > 0,$$

which is assured by the production function characteristics

$$\text{(B24)} \qquad F_{NR} > 0, \qquad F_{NN} < 0, \qquad F_{RR} < 0.$$

Standard analysis of Appendix B2 yields the comparative statics equations

$$\text{(B25)} \qquad N_w^{DCD} = (F_R)^2/D_2 < 0$$

$$N_v^{DCD} = -w(F_R)^2/vD_2 > 0$$

$$R_w^{DCD} = -F_N F_R/D_2 > 0$$

$$R_v^{DCD} = w\ F_N F_R/vD_2 < 0$$

$$N_{\bar{Y}}^{DCD} = -(v\ F_{NR} - w\ F_{RR})/D_2 > 0$$

$$R_{\bar{Y}}^{DCD} = -(w\ F_{NR} - v\ F_{NN})/D_2 > 0$$

$$D_2 = -(v\ F_N\ F_{NR} + w\ F_R\ F_{NR} - v\ F_R\ F_{NN} - w\ F_N\ F_{RR}) < 0.$$

Note that the DC case can only occur when

$$\text{(B26)} \qquad \bar{Y} < F(\bar{N}^S,\ \bar{R}^S)$$

since output is limited by the supply of factors.

(vi) With $\lambda_2 = 0$, $\lambda_3 > 0$, $\lambda_4 > 0$, the firm is constrained on the output and resource markets, but unconstrained on the labour market. This is the DRC case. Solution of the conditions yields

$$\text{(B27)} \qquad N = N^{DRCD} < \bar{N}^S$$

$$R = \bar{R}^S < R^{DCD}$$

$$Y = F(N,\ R) = \bar{Y} < Y^{RCS}$$

with

$$\text{(B28)} \qquad F_N(N,\ R) = w/\lambda_1$$

$$F_R(N,\ R) > v/\lambda_1, \qquad 0 < \lambda_1 \le 1,$$

which can be solved to obtain N in terms of $\bar{R}^S$ and $\bar{Y}$:

$$\text{(B29)} \qquad N = N^{DRCD}(\bar{Y}, \bar{R}^S)$$

$$R = \bar{R}^S$$

$$Y = F(N, R) = \bar{Y}.$$

Standard analysis yields the comparative statics equations

$$\text{(B30)} \qquad N_w^{DRCD} = 0$$

$$N_v^{DRCD} = 0.$$

Note that the case DRC can only occur when

$$\text{(B31)} \qquad \bar{Y} < F(\bar{N}^S, \bar{R}^S).$$

(vii) With $\lambda_3 = 0$, $\lambda_2 > 0$, $\lambda_4 > 0$, the firm is constrained on the output and labour markets, but unconstrained on the resource market. This is the DNC case. Solution of the conditions yields

$$\text{(B32)} \qquad N = \bar{N}^S < N^{DCD}$$

$$R = R^{DNCD} < \bar{R}^S$$

$$Y = F(N, R) = \bar{Y} < Y^{NCS}$$

with

$$\text{(B33)} \qquad F_N(N, R) > w/\lambda_1$$

$$F_R(N, R) = v/\lambda_1, \qquad 0 < \lambda_1 \le 1,$$

which can be solved to obtain R in terms of $\bar{N}^S$ and $\bar{Y}$

(B34)
$$N = \bar{N}^S$$
$$R = R^{DNCD}(\bar{Y}, \bar{N}^S)$$
$$Y = F(N, R) = \bar{Y}.$$

Standard analysis yields the comparative statics equations

(B35)
$$R_w^{DNCD} = 0$$
$$R_v^{DNCD} = 0.$$

Note that the DNC case can only occur when

(B36)
$$\bar{Y} < F(\bar{N}^S, \bar{R}^S)$$

(viii) With $\lambda_2 > 0$, $\lambda_3 > 0$, $\lambda_4 > 0$, the firm is constrained on all three markets. This is the DNRC case. Solution of the conditions yields

(B37)
$$N = \bar{N}^S \leq N^{DRCD}$$
$$R = \bar{R}^S \leq R^{DNCD}$$
$$Y = F(N, R) = \bar{Y} \leq Y^{NRCD}$$

with

(B38)
$$F_N = (w + \lambda_2)/\lambda_1$$
$$F_R = (v + \lambda_3)/\lambda_1, \qquad 0 < \lambda_1 \leq 1.$$

But this system of equations is over-determined. Only if autonomous demand equals full-employment output

(B39)
$$\bar{Y} = F(\bar{N}^S, \bar{R}^S)$$

will the system be consistent. But in that case we can think of the autonomous output as not constraining the representative firm, which is constrained by the supplies of factor inputs. Then $\lambda_1 = 1$ and both marginal products are greater than their respective real prices. But the region DNRC can only occur when

$$\bar{Y} = F(\bar{N}^S, \bar{R}^S),$$

and so there is general-market-clearing throughout region DNRC, with

(B40) $$\bar{Y} = Y^{NRCS} = F(\bar{N}^S, \bar{R}^S),$$

$$\bar{N}^S = Y^{DRCD}(\bar{Y}, \bar{R}^S),$$

$$\bar{R}^S = R^{DNCD}(\bar{Y}, \bar{N}^S).$$

If

(B41) $$\bar{Y} > F(\bar{N}^S, \bar{R}^S)$$

we have case NRC, and if

(B42) $$\bar{Y} < F(\bar{N}^S, \bar{R}^S)$$

we have case DC.

B2: Appendix to section 3.4.2.

In the DC case the first-order conditions from Appendix B1 can be written

$$\text{(B43)} \qquad \bar{Y} - F(N^{DCD}(w/v,\ \bar{Y}),\ R^{DCD}(w/v,\ \bar{Y})) \ = \ 0$$

$$F_N(N^{DCD},\ R^{DCD}) - F_R(N^{DCD},\ R^{DCD})\cdot w/v \ = \ 0.$$

To obtain the partial derivatives of the derived functions N^{DCD} and R^{DCD}, we differentiate these conditions with respect to $\bar{Y}$ and ρ, where ρ is the ratio of the real prices,

$$\text{(B44)} \qquad \rho \ \equiv \ w/v.$$

In matrix form we obtain

$$\text{(B45)} \qquad \begin{bmatrix} -F_N & -F_R \\ F_{NN} - \rho F_{NR} & F_{NR} - \rho F_{RR} \end{bmatrix} \begin{bmatrix} N^{DCD}_{\bar{Y}} & N^{DCD}_{\rho} \\ R^{DCD}_{\bar{Y}} & R^{DCD}_{\rho} \end{bmatrix} = \begin{bmatrix} -1 & 0 \\ 0 & F_R \end{bmatrix}$$

$$\therefore \quad \begin{bmatrix} N^{DCD}_{\bar{Y}} & N^{DCD}_{\rho} \\ R^{DCD}_{\bar{Y}} & R^{DCD}_{\rho} \end{bmatrix} = \frac{v}{D_2} \begin{bmatrix} F_{NR} - \rho F_{RR} & F_R \\ -F_{NN} + \rho F_{NR} & -F_N \end{bmatrix} \begin{bmatrix} -1 & 0 \\ 0 & F_R \end{bmatrix}$$

where

$$\text{(B46)} \qquad D_2 \ \equiv \ -(vF_NF_{NR} + wF_RF_{NR} - vF_RF_{NN} - wF_NF_{RR}) < 0.$$

This leads to

$$\text{(B47)} \qquad N^{DCD}_{\bar{Y}} \ = \ (wF_{RR} - vF_{NR})/D_2 > 0$$

(B47) $$R_{\bar{Y}}^{DCD} = (vF_{NN} - wF_{NR})/D_2 > 0$$

$$N_{\rho}^{DCD} = v(F_R)^2/D_2 < 0$$

$$R_{\rho}^{DCD} = -vF_N F_R/D_2 > 0.$$

Given that

(B48) $$\frac{\partial\rho}{\partial v} = -w/v^2 \quad \text{and} \quad \frac{\partial\rho}{\partial w} = 1/v,$$

we obtain the partial derivatives shown in Appendix B1, equation (B25):

(B49) $$N_{w}^{DCD} = (F_R)^2/D_2 < 0,$$

$$N_{v}^{DCD} = -w(F_R)^2/v\ D_2 > 0,$$

$$R_{w}^{DCD} = -F_N F_R/D_2 > 0,$$

$$R_{v}^{DCD} = wF_N F_R/v\ D_2 < 0.$$

B3: Appendix to section 3.4.4.

Consider the production function for neutral technical change

(B50) $$F^2(N, R) \equiv \alpha F(N, R), \qquad \alpha > 1.$$

The first-order conditions for case DC can be written

(B51) $$\bar{Y} - \alpha F(N, R) = 0$$

$$\alpha v F_N - \alpha w F_R = 0.$$

Differentiating totally with respect to α, N, and R we obtain

(B52) $$\begin{bmatrix} \alpha F_N & \alpha F_R \\ vF_{NN} - wF_{NR} & vF_{NR} - wF_{RR} \end{bmatrix} \begin{bmatrix} dN \\ dR \end{bmatrix} = -d\alpha \begin{bmatrix} F(N, R) \\ 0 \end{bmatrix}$$

$$\therefore \begin{bmatrix} dN \\ dR \end{bmatrix} = +d\alpha \frac{1}{\alpha D_2} \begin{bmatrix} vF_{NR} - wF_{RR} & -\alpha F_R \\ wF_{NR} - vF_{NN} & \alpha F_N \end{bmatrix} \begin{bmatrix} F(N, R) \\ 0 \end{bmatrix}$$

(B53) $$\therefore dN = (vF_{NR} - wF_{RR})Fd\alpha/\alpha D_2 < 0$$

$$dR = (wF_{NR} - vF_{NN})Fd\alpha/\alpha D_2 < 0.$$

Consider the production function for resource-augmenting technical change

(B54) $$F^1(N, R) \equiv F(N, \alpha R), \qquad \alpha > 1.$$

The first-order conditions for case DC can be written

$$\text{(B55)} \qquad \bar{Y} - F(N, \alpha R) = 0$$

$$vF_N - \alpha wF_R = 0.$$

Differentiating then totally with respect to α, N, and R we obtain

$$\text{(B56)} \quad \begin{bmatrix} F_N & \alpha F_R \\ \alpha wF_{NR} - vF_{NN} & \alpha^2 wF_{RR} - \alpha vF_{NR} \end{bmatrix} \begin{bmatrix} dN \\ dR \end{bmatrix} = -d\alpha \begin{bmatrix} F_R R \\ \alpha wF_{RR}R + wF_R - vF_{NR}R \end{bmatrix}$$

$$\therefore \quad \begin{bmatrix} dN \\ dR \end{bmatrix} = -d\alpha \, \frac{1}{\alpha D_3} \begin{bmatrix} \alpha^2 wF_{RR} - \alpha vF_{NR} & -\alpha F_R \\ -\alpha wF_{NR} - vF_{NN} & F_N \end{bmatrix} \begin{bmatrix} F_R R \\ \alpha wF_{RR}R + wF_R - vF_{NR}R \end{bmatrix}$$

where

$$D_3 \equiv \alpha w(F_N F_{RR} - F_R F_{NR}) + v(F_R F_{NN} - F_N F_{NR}) < 0$$

$$\text{(B57)} \quad \therefore \; dN = w(F_R)^2 \, d\alpha / D_3 < 0$$

$$\text{(B58)} \quad \therefore \; dR = -(\alpha wR(F_N F_{RR} - F_R F_{NR}) + vR(F_R F_{NN} - F_N F_{NR}) + wF_N F_R) \, d\alpha / D_3$$

$$< 0 \quad \text{if } \alpha \text{ large}.$$

Thus we see that both in the case of neutral technical change and in the case of resource-augmenting technical change the level of employment N^{DCD} drops, as does the level of resource use R^{DCD}, except for α only just greater than one in the case of resource-augmenting technical change, in which case the level of resource use rises.

B4: Appendix to section 3.6.1.

Consider the resource-market-clearing locus:

(B59) $$R^D = \bar{R}^S.$$

Along R—DR, R—DNR, and R—NR of Figures 3.3a, 3.4, and 3.3b, respectively, R^D is $R^{SCD}(w, v)$. Total differentiation gives

(B60) $$\left.\frac{dw}{dv}\right|_{locus} = -R_v^{SCD}/R_w^{SCD} < 0.$$

Along DR—O of Figure 3.3a, R^D is $R^{DCD}(w/v, \bar{Y})$. Total differentiation gives

(B61) $$\left.\frac{dw}{dv}\right|_{locus} = -\frac{\partial\rho}{\partial v} / \frac{\partial\rho}{\partial w} = w/v,$$

which means that DR—O is a straight line. Along NR—R' of Figure 3.3b, R^D is $R^{NCD}(v, \bar{N}^S)$. Total differentiation gives

(B62) $$\left.\frac{dw}{dv}\right|_{locus} = \infty.$$

The labour-market-clearing locus can be analysed similarly.

Consider the output-market-clearing locus:

(B63) $$Y^S = \bar{Y} = Y^D.$$

Along DR—DN of Figure 3.3a, Y^S is $Y^{SCS}(w, v)$. Total differentiation gives

(B64) $$\left.\frac{dw}{dv}\right|_{locus} = -Y_v^{SCS}/Y_w^{SCS} < 0.$$

Along Y—DR and YN—DNR of Figures 3.3a and 3.4, Y^S is $Y^{RCS}(w, \bar{R}^S)$. Total differentiation gives

(B65) $$\left.\frac{dw}{dv}\right|_{\text{locus}} = 0.$$

Along Y—DN and YR—DNR of Figures 3.3a and 3.4, Y^S is $Y^{NCS}(v, \bar{N}^S)$. Total differentiation gives

(B66) $$\left.\frac{dw}{dv}\right|_{\text{locus}} = \infty.$$

Hence the figures as shown.

Following Appendix A2, if the production function is Cobb-Douglas with diminishing returns to scale

$$Y = F(N, R) = N^{\beta}R^{\gamma}; \qquad \beta+\gamma=1-\alpha;\ \alpha,\beta,\gamma>0,$$

then the slopes of three market-clearing loci bounding the SC region in Figures 3.3a, 3.3b, and 3.4 are given by (A15),

(B67) $$\left.\frac{dw}{dv}\right|_{N^X} = -\beta R/(\alpha+\beta)N,$$

$$\left.\frac{dw}{dv}\right|_{R^X} = -(\alpha+\gamma)R/\gamma N,$$

$$\left.\frac{dw}{dv}\right|_{Y^X} = -R/N,$$

whence we see that for any R and N

(B68) $$\left.\frac{dw}{dv}\right|_{R^X} < \left.\frac{dw}{dv}\right|_{Y^X} < \left.\frac{dw}{dv}\right|_{N^X} < 0$$

as shown in the figures.

B5: <u>Appendix to section 3.6.2.</u>

Consider neutral technical change

(B69) $$F^2(N, R) \equiv \alpha F(N, R), \qquad \alpha > 1.$$

In region RC the first-order conditions can be written

(B70) $$\alpha F_N(N, \bar{R}^S) = w.$$

Differentiating with respect to α and N, we get

(B71) $$F_N \, d\alpha + \alpha F_{NN} \, dN = 0$$

$$\therefore \quad dN = -F_N \, d\alpha/\alpha F_{NN} > 0,$$

so that the demand for labour NRCD increases. The output in this region is given by

(B72) $$Y = \alpha F(N, \bar{R}^S).$$

Differentiating, we obtain

(B73) $$dY = F \, d\alpha + \alpha F_N \, dN > 0.$$

The supply of output Y^{RCS} increases.

Symmetrical analysis shows that in region NC, neutral technical change results in increased demand for resource flow R^{NCD} and increased supply of output Y^{NCS}.

Consider resource-augmenting technical change

$$F^1(N, R) \equiv F(N, \alpha R), \qquad \alpha > 1.$$

In region RC the first-order conditions can be written

(B75) $$F_N(N, \alpha\bar{R}^S) = w.$$

Differentiating with respect to α and N, we get

(B76) $$F_{NR}\,\bar{R}^S\,d\alpha + F_{NN}\,dN = 0$$

$$\therefore\quad dN = -F_{NR}\,\bar{R}^S\,d\alpha/F_{NN} > 0$$

so that the demand for labour N^{RCD} increases. The output Y^{RCS} also increases.

In region NC the first-order conditions can be written

(B77) $$\alpha F_R(\bar{N}^S, \alpha R) = v.$$

Differentiating with respect to α and R, we get

(B78) $$F_R\,d\alpha + \alpha^2 F_{RR}\,dR + \alpha RF_{RR})\,d\alpha = 0$$

$$\therefore\quad dR = -(F_R + \alpha RF_{RR})\,d\alpha/\alpha^2 F_{RR}$$

$$< 0 \quad \text{iff} \quad \alpha \text{ large.}$$

The demand for resource flow R^{NCD} falls for large α. The output for this region is given by

(B79) $$Y = F(\bar{N}^S, \alpha R).$$

Differentiating, we obtain

(B80) $$dY = F_R R\,d\alpha + F_R\,\alpha\,dR$$

$$= -F_R^2\,d\alpha/\alpha F_{RR} > 0.$$

The supply of output Y^{NCD} increases.

APPENDIX C: APPENDIX TO CHAPTER IV

C1: Appendix to section 4.1.1.

Consider the constant-v locus with the simple formulation (SF) in region DC:

$$\text{(C1)} \qquad \lambda_V(R^{DCD}(w/v, \bar{Y}) - \bar{R}^S) - \lambda_P(\bar{Y} - Y^{SCS}(w, v)) = 0.$$

Differentiating totally with respect to w and v, we get

$$\text{(C2)} \qquad \left.\frac{dw}{dv}\right|_v = -(\lambda_V R_v^{DCD} + \lambda_P Y_v^{SCS})/(\lambda_V R_w^{DCD} + \lambda_P Y_w^{SCS})$$

$$\gtrless 0 \text{ as } \lambda_V R_w^{DCD} \gtrless -\lambda_P Y_w^{SCS}.$$

Hence the slope of the constant-v locus depends on the relative magnitude of the speeds of adjustment, λ_V and λ_P. A similar expression can be obtained for the slope of the constant-w locus.

Consider the constant-v locus with the simple formulation (SF) in region NRC:

$$\text{(C3)} \qquad \lambda_V(R^{NCD}(v, \bar{N}^S) - \bar{R}^S) - \lambda_P(Y^D - Y^{NRCS}(\bar{N}^S, \bar{R}^S)).$$

Differentiating totally with respect to w and v, we get that

$$\text{(C4)} \qquad \left.\frac{dw}{dv}\right|_v = \infty.$$

Hence the constant-v locus is vertical in region NRC. It is readily found that the constant-w locus is horizontal in NRC. Hence Q_{SF}^{NRC} is unique.

C2: Appendix to section 4.1.2.

The equations for the rates of change of W, V, and P in the alternative formulation (AF) are

$$\text{(C5)} \qquad \dot{W}/W = \lambda_W N^X + m\cdot\dot{P}/P, \qquad 0 < m \leq 1,$$

$$\dot{V}/V = \lambda_V R^X + n\cdot\dot{P}/P, \qquad 0 \leq n \leq 1,$$

$$\dot{P}/P = \lambda_P Y^X + j\cdot\dot{W}/W + k\cdot\dot{V}/V, \qquad 0 \leq (j+k) < 1;\ j,k > 0.$$

Elimination of $\dot{P}/P$ leads to

$$\text{(C6)} \qquad g \equiv \dot{v}/v = \lambda_V^V R^X - \lambda_P^V Y^X - \lambda_W^V N^X,$$

$$h \equiv \dot{w}/w = \lambda_W^W N^X - \lambda_P^W Y^X - \lambda_V^W R^X,$$

where

$$\text{(C7)} \qquad \lambda_V^V \quad \lambda_V(1 - jm - k)/(1 - jm - kn)$$

$$\lambda_P^V = \lambda_P(1-n)/(1 - jm - kn)$$

$$\lambda_W^V = \lambda_W j(1-n)/(1 - jm - kn)$$

$$\lambda_W^W = \lambda_W(1 - j - kn)/(1 - jm - kn)$$

$$\lambda_P^W = \lambda_P(1-m)/(1 - jm - kn)$$

$$\lambda_V^W = \lambda_V k(1-m)/(1 - jm - kn).$$

Consider the constant-v locus with the alternative formulation (AF) in region NRC:

(C8) $$\lambda_V^V(R^{NCD} - \bar{R}^S) - \lambda_P^V(Y^D - Y^{NRCS}) - \lambda_W^V(N^{RCD} - \bar{N}^S) = 0.$$

Total differentiation with respect to w and v leads to

(C9) $$\left.\frac{dw}{dv}\right|_v = \lambda_V^V R_v^{NCD}/\lambda_W^V N_w^{RCD} > 0.$$

Hence the constant-v locus is positively sloped in region NRC. Similar analysis shows that the constant-w locus is positively sloped in region NRC as well.

C3: Appendix to section 4.2.1.

Consider the equations of motion

(C10) $$\dot{v}/v = g(w, v)$$

$$\dot{w}/w = h(w, v).$$

We can examine the local dynamic stability of this system around any point (w*, v*) by forming the characteristic equation of the system

(C11) $$\begin{vmatrix} v^*g_v - Z & v^*g_w \\ w^*h_v & w^*h_w - Z \end{vmatrix} = 0$$

where the partial derivatives of g and h with respect to the two state variables w and v are evaluated in the neighbourhood of the point (w*, v*). The system will exhibit local dynamic stability in the region of the poing (w*, v*) iff the real parts of the roots of the characteristic equation are negative, Re(Z) < 0. It is necessary and sufficient for this that the trace of the Jacobian matrix J of the partial derivatives of $\dot{w}/w$ and $\dot{v}/v$ is negative and that its determinant is positive. That is, that B > 0 and C > 0, where

(C12) $$B = -(v^*g_v + w^*h_w),$$

$$C = w^*v^*(g_vh_w - g_wh_v) = w^*v^*|J|.$$

Furthermore, if $B^2 - 4C$ is negative, the equilibrium is a focus (there is cycling), and if $B^2 - 4C$ is positive, the equilibrium is a node (there is no cycling). These relationships are illustrated in Figure C1.

Consider the simple formulation of price adjustments:

(C13) $$g = \lambda_V R^X - \lambda_P Y^X,$$

$$h = \lambda_W N^X - \lambda_P Y^X.$$

Manipulation yields

(C14) $$B = -(\lambda_V v^* R_v^X + \lambda_W w^* N_w^X - \lambda_P(v^* Y_v^X + w^* Y_w^X))$$

and

(C15) $$C/w^*v^* = \lambda_V\lambda_W(R_v^X N_w^X - R_w^X N_v^X)$$

$$- \lambda_V\lambda_P(R_v^X Y_w^X - R_w^X Y_v^X)$$

$$- \lambda_W\lambda_P(N_w^X Y_v^X - N_v^X Y_w^X).$$

(i) Consider Q_{SF}^{DC}. The three excess effective demands are given by

(C16) $$R^X = R^{DCD} - \bar{R}^S < 0,$$

$$N^X = N^{DCD} - \bar{N}^S < 0,$$

$$Y^X = \bar{Y} - Y^{SCS} < 0.$$

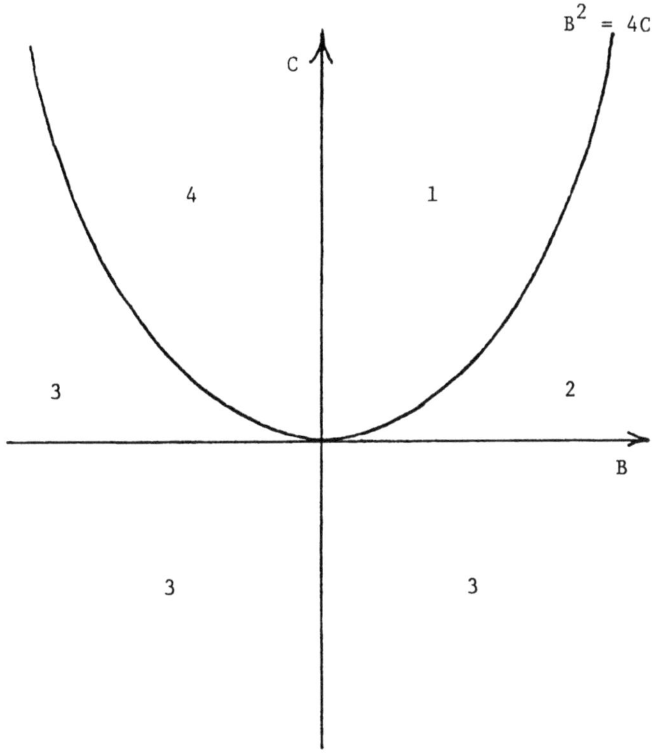

region 1: stable focus - convergence with spiralling

region 2: stable node - convergence without spiralling

region 3: unstable node - divergence without spiralling

region 4: unstable focus - divergence with spiralling

Figure C1: Typology of equilibria.

From Table 4.3, the signs of the elasticities of these excess effective demands are:

$$\text{positive:} \quad Y^X_v,\ Y^X_w,\ R^X_w,\ N^X_v,$$

$$\text{negative:} \quad R^X_v,\ N^X_w.$$

Thus we see by inspection that B is positive.

From equations (B25) we see that

$$\text{(C17)} \qquad R^X_v N^X_w = R^X_w N^X_v = wF_N(F_R)^3/v(D_2)^2.$$

Thus we can derive C:

$$\text{(C18)} \qquad C = -w^*v^* \lambda_P(\lambda_V(R^X_v Y^X_w - R^X_w Y^X_v) + \lambda_W(N^X_w Y^X_v - N^X_v Y^X_w)),$$

which is positive. Since both B and C are positive, the quasi-equilibrium Q^{DC}_{SF} is locally stable.

Consider the Cobb-Douglas production function

$$\text{(C19)} \qquad F(N, R) = N^\beta R^\gamma, \qquad \beta+\gamma=1-\alpha;\ \alpha,\beta,\gamma > 0.$$

We see that since $\beta+\gamma < 1$ this production function exhibits decreasing returns to scale. Substituting, we obtain

$$\text{(C20)} \qquad B^2 - 4C = (\lambda_V\beta\ R^{DCD}/(1-\alpha) + \lambda_W\gamma\ N^{DCD}/(1-\alpha) - \lambda_P(1-\alpha)\ Y^{SCS}/\alpha)^2,$$

which is positive: if the production function is Cobb-Douglas with decreasing returns to scale, quasi-equilibrium Q^{DC}_{SF} is a stable node.

(ii) Consider Q_{SF}^{NRC}. The three excess demands are given by

$$R^X = R^{NCD} - \bar{R}^S > 0, \tag{C21}$$

$$N^X = N^{RCD} - \bar{N}^S > 0,$$

$$Y^X = s(\bar{Y} - Y^{NRCS}) > 0.$$

From Table 4.3, the signs of the elasticities of these effective demands are:

zero: Y_v^X, Y_w^X, R_w^X, N_v^X

negative: R_v^X, N_w^X.

Thus we see by inspection that B is positive, and that

$$C = w^* v^* \lambda_V \lambda_W R_v^X N_w^X, \tag{C22}$$

which is positive. Since both B and C are positive, the quasi-equilibrium Q_{SF}^{NRC} is locally stable. Furthermore, manipulation yields

$$B^2 - 4C = (\lambda_V v^* R_v^X - \lambda_W w^* N_w^X)^2, \tag{C23}$$

which is positive: the quasi-equilibrium Q_{SF}^{NRC} is a stable node.

C4: Appendix to section 4.2.2.

Consider the alternative formulation of price adjustments:

$$g = \lambda_V^V R^X - \lambda_P^V Y^X - \lambda_W^V N^X, \tag{C24}$$

$$h = \lambda_W^W N^X - \lambda_P^W Y^X - \lambda_V^W R^X.$$

Manipulation yields values for B and C of the previous appendix:

$$\text{(C25)} \qquad B = -(\lambda_V^V v^* R_v^X - \lambda_V^W w^* R_w^X + \lambda_W^W w^* N_w^X - \lambda_W^V v^* N_v^X - \lambda_P^V v^* Y_v^X - \lambda_P^W w^* Y_w^X),$$

and

$$\text{(C26)} \qquad C(1 - jm - kn)/w^*v^* = \lambda_V \lambda_W(1 - (j+k))(R_v^X N_w^X - R_w^X N_v^X) - \lambda_V \lambda_P(1-m)(R_v^X Y_w^X - R_w^X Y_v^X) - \lambda_W \lambda_P(1-n)(N_w^X Y_v^X - N_v^X Y_w^X).$$

(i) Consider Q_{AF}^{SC}. The three excess effective demands are given by

$$\text{(C27)} \qquad R^X = R^{SCD} - \bar{R}^S < 0,$$

$$N^X = N^{SCD} - \bar{N}^S < 0,$$

$$Y^X = s(\bar{Y} - Y^{SCS}) > 0.$$

From Table 4.3, the signs of the elasticities of these excess effective demands are

positive: Y_v^X, Y_w^X

negative: R_v^X, R_w^X, N_v^X, N_w^X.

Manipulation of the expression for B yields the condition that B is positive iff

$$\text{(C28)} \qquad F_{NR}(w^*v^*s(\lambda_P^V + \lambda_P^W) - v^*\lambda_W^V - w^*\lambda_V^W)$$

$$- F_{RR}\ w^*(\lambda_W^W + w^*s\lambda_P^W)$$

$$- F_{NN}\ v^*(\lambda_V^V + v^*s\lambda_P^V)$$

is positive. Thus we can see that $F_{NR} = 0$ is sufficient for B to be positive. Manipulation yields

$$\text{(C29)} \qquad C = w^*v^*(v^*s\ \lambda_W\lambda_P(1-n) + w^*s\ \lambda_V\lambda_P(1-m)$$

$$+ \lambda_W\lambda_V(1 - (j+k)))/D_1(1-jm-kn),$$

which can be seen to be positive. Thus we see that it is sufficient for the quasi-equilibrium Q_{AF}^{SC} to be locally stable that F_{NR} is almost zero.

(ii) Consider Q_{AF}^{DC}. The three excess effective demands are given by

$$\text{(C30)} \qquad R^X = R^{DCD} - \bar{R}^S < 0,$$

$$N^X = N^{DCD} - \bar{N}^S < 0,$$

$$Y^X = \bar{Y} - Y^{SCS} < 0.$$

From Table 4.3, the signs of the elasticities of these excess effective demands are:

positive: Y_v^X, Y_w^X, R_w^X, N_v^X

negative: R_v^X, N_w^X.

Thus we see by inspection that B is positive. Since

(C31) $$R^X_v N^X_w = R^X_w N^X_v,$$

as shown in the previous appendix section, we can derive C

(C32) $$C = -w^*v^*(\lambda_V \lambda^W_P(R^X_v Y^X_w - R^X_w Y^X_v) + \lambda_W \lambda^V_P(N^X_w Y^X_v - N^X_v Y^X_w)),$$

which can be seen to be positive. Since B and C are positive, the quasi-equilibrium Q^{DC}_{AF} is locally stable.

(iii) Consider Q^{NRC}_{AF}. The three excess effective demands are given by

(C33) $$R^X = R^{NCD} - \bar{R}^S > 0,$$

$$N^X = N^{RCD} - \bar{N}^S > 0,$$

$$Y^X = s(\bar{Y} - Y^{NRCS}) > 0.$$

From Table 4.3, the signs of the elasticities of these excess effective demands are:

zero: Y^X_v, Y^X_w, R^X_w, N^X_v,

negative: R^X_v, N^X_w.

Thus we see by inspection that B is positive, and that

(C34) $$C = w^*v^* \lambda_V \lambda_W(1 - (j+k)) R^X_v N^X_w/(1-jm-kn),$$

which is positive. Since both B and C are positive, the quasi-equilibrium Q^{NRC}_{AF} is locally stable. Manipulation yields

(C35) $$B^2 - 4C = (\lambda^V_V v^* R^X_v - \lambda^W_W w^* N^X_w)^2 + 4\lambda^W_V \lambda^V_W w^*v^* R^X_v N^X_w,$$

which is positive: the quasi-equilibrium Q^{NRC}_{AF} is a stable node.

C5: Appendix to section 4.3.1.

Consider the equations defining the point of quasi-equilibrium

(C36) $$\dot{v}/v = g(w^*, v^*) = 0$$

$$\dot{w}/w = h(w^*, v^*) = 0.$$

Differentiating totally with respect to w, v, R^X, N^X, and Y^X, we obtain

(C37) $$\begin{bmatrix} g_v & g_w \\ h_v & h_w \end{bmatrix} \begin{bmatrix} dv^* \\ dw^* \end{bmatrix} = - \begin{bmatrix} g_{R^X} & g_{N^X} & g_{Y^X} \\ h_{R^X} & h_{N^X} & h_{Y^X} \end{bmatrix} \begin{bmatrix} dR^X \\ dN^X \\ dY^X \end{bmatrix}.$$

We designate as J the Jacobian matrix of the partial derivatives of $\dot{v}/v$ and $\dot{w}/w$. We note that

(C38) $$|J| = C/w^*v^*,$$

where C is the determinant of the characteristic equation, considered in the stability analyses of Appendices C3 and C4.

Then the expression can be written

(C39) $$\begin{bmatrix} dv^* \\ dw^* \end{bmatrix} = - J^{-1} \begin{bmatrix} g_{R^X} & g_{N^X} & g_{Y^X} \\ h_{R^X} & h_{N^X} & h_{Y^X} \end{bmatrix} \begin{bmatrix} dR^X \\ dN^X \\ dY^X \end{bmatrix}$$

where

$$\text{(C40)} \qquad J^{-1} = \frac{1}{|J|}\begin{bmatrix} h_w & -g_w \\ -h_v & g_v \end{bmatrix},$$

and we note that $|J|$ is positive if the quasi-equilibrium is locally stable at (w*, v*).

For the simple formulation of price adjustment, we have

$$\text{(C41)} \qquad g = \lambda_V R^X - \lambda_P Y^X$$

$$h = \lambda_W N^X - \lambda_P Y^X,$$

from equations (4.7). Thus equation (C39) can be written

$$\text{(C42)} \qquad \begin{bmatrix} dv^* \\ dw^* \end{bmatrix} = -\frac{1}{|J|}\begin{bmatrix} h_w & -g_w \\ -h_v & g_v \end{bmatrix}\begin{bmatrix} \lambda_V & 0 & -\lambda_P \\ 0 & \lambda_W & -\lambda_P \end{bmatrix}\begin{bmatrix} dR^X \\ dN^X \\ dY^X \end{bmatrix}$$

$$= -\frac{1}{|J|}\begin{bmatrix} h_w & -g_w \\ -h_v & g_v \end{bmatrix}\begin{bmatrix} \lambda_V dR^X & 0 & -\lambda_P dY^X \\ 0 & \lambda_W dN^X & -\lambda_P dY^X \end{bmatrix}.$$

We can now proceed with the comparative statics.

1. Consider quasi-equilibrium Q_{SF}^{DC}. The three excess effective demands, from Table 4.1, are

$$\text{(C43)} \qquad R^X = R^{DCD}(w/v, \bar{Y}) - \bar{R}^S < 0,$$

$$N^X = N^{DCD}(w/v, \bar{Y}) - \bar{N}^S < 0,$$

$$Y^X = \bar{Y} - Y^{SCS}(w,v) < 0.$$

Then

$$\text{(C44)} \qquad J^{-1} = \frac{1}{|J|}\begin{bmatrix} \lambda_W N_w^{DCD} + \lambda_P Y_w^{SCS} & -\lambda_V R_w^{DCD} - \lambda_P Y_w^{SCS} \\ -\lambda_W N_v^{DCD} - \lambda_P Y_v^{SCS} & \lambda_V R_v^{DCD} + \lambda_P Y_v^{SCS} \end{bmatrix}$$

where $|J|$ was shown to be positive in equation (C18).

1a. Consider an increase in resource supply, $d\bar{R}^S$ positive. Then

$$\text{(C45)} \qquad dR^X = -d\bar{R}^S < 0,$$

$$dN^X = 0,$$

$$dY^X = 0.$$

Solving for the two endogenous variables, we obtain

$$\text{(C46)} \qquad dv^* = (\lambda_W N_w^{DCD} + \lambda_P Y_w^{SCS}) \lambda_V d\bar{R}^S/|J| < 0,$$

$$dw^* = -(\lambda_W N_v^{DCD} + \lambda_P Y_v^{SCS}) \lambda_V d\bar{R}^S/|J| \gtrless 0.$$

We conclude that v* will fall, while w* may fall or rise.

1b. Consider an increase in autonomous demand for output, $d\bar{Y}$ positive. Then

$$\text{(C47)} \qquad dR^X = R_{\bar{Y}}^{DCD} d\bar{Y} > 0,$$

$$dN^X = N_{\bar{Y}}^{DCD} d\bar{Y} > 0,$$

$$dY^X = d\bar{Y} > 0.$$

Solving for the two endogenous variables, we obtain

$$\text{(C48)} \qquad dv^* = -((\lambda_W N_w^{DCD} + \lambda_P Y_w^{SCS})\lambda_V R_{\bar{Y}}^{DCD} - (\lambda_V R_w^{DCD} + \lambda_P Y_w^{SCS})\lambda_W N_{\bar{Y}}^{DCD} - \lambda_P(\lambda_W N_w^{DCD} - \lambda_V R_w^{DCD}))\, d\bar{Y}/|J|.$$

Examination of the signs of the components of this expression shows that v* may rise or fall, depending on the relative magnitudes of the exogenous variables. A symmetrical argument leads to the same concludion for w*.

1c. Consider neutral technical change,

$$\text{(C49)} \qquad F^2(N, R) \equiv \alpha F(N, R), \qquad \alpha > 1,$$

with $d\alpha$ positive. Then, from Appendix B3, equations (B53),

$$\text{(C50)} \qquad dR^X = (wF_{NR} - vF_{NN})\, Fd\alpha/\alpha D_2 < 0$$

$$dN^X = (vF_{NR} - wF_{RR})\, Fd\alpha/\alpha D_2 < 0$$

$$dY^X > 0.$$

If we perform the tedious manipulations for calculation of dv* and dw* both for neutral and for resource-augmenting technical change, we find that the signs of dv* and dw* are inconclusive, depending on the relative magnitudes of the exogenous variables.

2. Consider quasi-equilibrium Q_{SF}^{NRC}. The three excess effective demands, from Table 4.1, are

$$\text{(C51)} \qquad R^X = R^{NCD}(v, \bar{N}^S) - \bar{R}^S > 0,$$

$$N^X = N^{RCD}(w, \bar{R}^S) - \bar{N}^S > 0,$$

$$Y^X = s(\bar{Y} - Y^{NRCS}(\bar{N}^S, \bar{R}^S)) > 0.$$

Then

$$\text{(C52)} \qquad J^{-1} = \frac{1}{|J|} \begin{bmatrix} \lambda_W \, N_w^{RCD} & 0 \\ 0 & \lambda_V \, R_v^{NCD} \end{bmatrix}$$

where $|J|$ was shown to be positive in equation (C22).

2a. Consider an increase in the resource supply, $d\bar{R}^S$ positive. Then

$$\text{(C53)} \qquad dR^X = -d\bar{R}^S < 0,$$

$$dN^X = N_{\bar{R}^S}^{RCD} \, d\bar{R}^S = -F_{NR} \, d\bar{R}^S / F_{NN} > 0,$$

$$dY^X = -s \, Y_{\bar{R}^S}^{NRCS} \, d\bar{R}^S = - s \, F_R \, d\bar{R}^S < 0.$$

Solving for the two endogenous variables, we obtain

$$\text{(C54)} \qquad dv^* = -\lambda_W \, N_w^{RCD}(-\lambda_V + s \, \lambda_P F_R) \, d\bar{R}^S / |J| \gtrless 0,$$

$$dw^* = -\lambda_V \, R_v^{NCD}(-\lambda_W \, F_{NR}/F_{NN} + s \, \lambda_P F_R) \, d\bar{R}^S / |J| > 0.$$

Thus we see that w* will rise, while v* may rise or fall.

2b. Consider an increase in autonomous demand for output, $d\bar{Y}$ positive. Then

$$\text{(C55)} \qquad dR^X = 0,$$

$$dN^X = 0,$$

$$dY^X = s \, d\bar{Y} > 0.$$

Solving for the two endogenous variables, we obtain

(C56) $$dv^* = s\ \lambda_W \lambda_P\ N_w^{RCD}\ dY^X/|J| < 0,$$

$$dw^* = s\ \lambda_V \lambda_P\ R_v^{NCD}\ dY^X/|J| < 0.$$

Thus we see that both w* and v* will fall with $d\bar{Y}$ positive.

2c. Consider neutral technical change,

(C57) $$F^2(N, R) \equiv \alpha F(N, R), \qquad \alpha > 1,$$

with $d\alpha$ positive. Then, from Appendix B5,

(C58) $$dR^X = -F_R\ d\alpha/\alpha\ F_{RR} > 0,$$

$$dN^X = -F_N\ d\alpha/\alpha\ F_{NN} > 0,$$

$$dY^X = -sF\ d\alpha < 0.$$

Solving for the two endogenous variables, we obtain

$$dv^* = \lambda_W\ N_w^{RCD}(\lambda_V F_R/\alpha\ F_{RR} - s\lambda_P F)d\alpha/|J| > 0,$$

$$dw^* = \lambda_V\ R_v^{NCD}(\lambda_W F_N/\alpha\ F_{NN} - s\lambda_P F)d\alpha/|J| > 0.$$

Thus we see that both w* and v* will rise with neutral technical change.

2d. Consider resource-augmenting technical change,

(C60) $$F^2(N, R) \equiv F(N, \alpha R), \qquad \alpha > 1,$$

with $d\alpha$ positive. Then, from Appendix B5,

(C61) $$dR^X = -(F_R + \alpha R^{NCD}\ F_{RR})\ d\alpha/\alpha^2\ F_{RR} \lesseqgtr 0,$$

$$dN^X = -F_{NR}\ \bar{R}^S\ d\alpha/F_{NN} > 0,$$

$$dY^X = -sF_R\ \bar{R}^S\ d\alpha < 0.$$

Solving for the two endogenous variables, we obtain

(C62) $$dv^* = \lambda_W N_w^{RCD}(\lambda_V(F_R + \alpha R^{NCD} F_{RR})/\alpha^2 F_{RR} - s\lambda_P F_R \bar{R}^S)d\alpha/|J| \gtrless 0$$

(C63) $$dw^* = \lambda_V R_v^{NCD}(\lambda_W F_{NR}/F_{NN} - s\lambda_P F_R)\bar{R}^S d\alpha/|J| > 0.$$

Thus we see that w* will always rise, and, that for large α, v* will rise with resource-augmenting technical change.

C6: Appendix to section 4.3.2.

For the alternative formulation of price adjustment, we have

(C64) $$g = \lambda_V^V R^X - \lambda_P^V Y^X - \lambda_W^V N^X,$$

$$h = \lambda_W^W N^X - \lambda_P^W Y^X - \lambda_V^W R^X,$$

from equations (4.18). Thus, from equation (C42), we can write

(C65) $$\begin{bmatrix} dv^* \\ dw^* \end{bmatrix} = \frac{-1}{|J|}\begin{bmatrix} h_w & -g_w \\ -h_v & g_v \end{bmatrix}\begin{bmatrix} \lambda_V^V & -\lambda_W^V & -\lambda_P^V \\ -\lambda_V^W & \lambda_W^W & -\lambda_P^W \end{bmatrix}\begin{bmatrix} dR^X \\ dN^X \\ dY^X \end{bmatrix}$$

$$= \frac{-1}{|J|}\begin{bmatrix} h_w & -g_w \\ -h_v & g_v \end{bmatrix}\begin{bmatrix} \lambda_V^V dR^X & -\lambda_W^V dN^X & -\lambda_P^V dY^X \\ -\lambda_V^W dR^X & \lambda_W^W dN^X & -\lambda_P^W dY^X \end{bmatrix}$$

We can now proceed with the comparative statics.

1. Consider quasi-equilibrium Q^{SC}_{AF}. The three excess demands, from Table 4.1, are

(C66) $$R^X = R^{SCD}(w, v) - \bar{R}^S < 0,$$

$$N^X = N^{SCD}(w, v) - \bar{N}^S < 0,$$

$$Y^X = s(\bar{Y} - Y^{SCS}(w, v)) > 0.$$

Then

(C67) $$J^{-1} = \frac{1}{|J|}\begin{bmatrix} \lambda^W_W N^{SCD}_w + s\lambda^W_P Y^{SCS}_w - \lambda^W_V R^{SCD}_w & -\lambda^V_V R^{SCD}_w - s\lambda^V_P Y^{SCS}_w + \lambda^V_W N^{SCD}_w \\ -\lambda^W_W N^{SCD}_v - s\lambda^W_P Y^{SCS}_v + \lambda^W_V R^{SCD}_v & \lambda^V_V R^{SCD}_v + s\lambda^V_P Y^{SCS}_v - \lambda^V_W N^{SCD}_v \end{bmatrix}$$

where $|J|$ is shown to be positive in equation (C29).

1a. Consider an increase in resource supply, $d\bar{R}^S$ positive. Then

(C68) $$dR^X = -d\bar{R}^S < 0,$$

$$dN^X = 0,$$

$$dY^X = 0.$$

Solving for the two endogenous variables, we obtain

(C69) $$dv^* = (\lambda^V_V(\lambda^W_W N^{SCD}_w + s\lambda^W_P Y^{SCS}_w - \lambda^W_V R^{SCD}_w) + \lambda^W_V(\lambda^V_V R^{SCD}_w + s\lambda^V_P Y^{SCS}_w - \lambda^V_W N^{SCD}_w))d\bar{R}^S/|J|$$

(C70) $$= \lambda_V(\lambda_W(1 - (j+k))N^{SCD}_w + s\lambda_P(1-m)Y^{SCS}_w)d\bar{R}^S/(1-jm-kn)|J|,$$

which is negative, and

(C71) $$dw^* = -(\lambda_V^V(\,{}_W^W N_v^{SCD} + s\lambda_P^W Y_v^{SCS} - \lambda_V^W R_v^{SCD})$$

$$+ \lambda_V^W(\lambda_V^V R_v^{SCD} + s\lambda_P^V Y_v^{SCS} - \lambda_W^V N_v^{SCD}))d\bar{R}^S/|J|$$

$$= -\lambda_V(\lambda_W(1-(j+k))N_v^{SCD} + s\lambda_P(1-m)Y_v^{SCS})d\bar{R}^S/(1-jm-kn)|J|,$$

which is positive. We conclude that v* will fall and w* will rise in response to an increase in $\bar{R}^S$.

1b. Consider an increase in autonomous demand for output, $d\bar{Y}$ positive. Then

(C72) $$dR^X = 0,$$

$$dN^X = 0,$$

$$dY^X = sd\bar{Y} > 0.$$

Solving for the two endogenous variables, we obtain

(C73) $$dv^* = s(\lambda_P^V(\lambda_W^W N_w^{SCD} + s\lambda_P^W Y_w^{SCS} - \lambda_V^W R_w^{SCD})$$

$$- \lambda_P^W(\lambda_V^V R_w^{SCD} + s\lambda_P^V Y_w^{SCS} - \lambda_W^V N_w^{SCD}))d\bar{Y}/|J|$$

$$= s\lambda_P(\lambda_W(1-n)N_w^{SCD} - \lambda_V(1-m)R_w^{SCD})d\bar{Y}/(1-jm-kn)|J|$$

$$\gtrless 0.$$

A symmetrical argument shows that w* will rise or fall, as well, depending on the magnitudes of the exogenous parameters.

1c. If we perform the tedious calculations to determine the signs of dv* and dw* for technical change (both neutral and resource-augmenting), we find that the signs are inconclusive, depending on the exogenous variables' relative magnitudes.

2. Consider the quasi-equilibrium Q_{AF}^{DC}. The three excess demands, from Table 4.1, are

$$\text{(C74)} \qquad \begin{aligned} R^X &= R^{DCD}(w/v,\ \bar{Y}) - \bar{R}^S < 0, \\ N^X &= N^{DCD}(w/v,\ \bar{Y}) - \bar{N}^S < 0, \\ Y^X &= \bar{Y} - Y^{SCS}(w,\ v) < 0. \end{aligned}$$

Then

$$\text{(C75)} \qquad J^{-1} = \frac{1}{|J|}\begin{bmatrix} \lambda_W^W N_w^{DCD} + \lambda_P^W Y_w^{SCS} - \lambda_V^W R_w^{DCD} & -\lambda_V^V R_w^{DCD} - \lambda_P^V Y_w^{SCS} + \lambda_W^V N_w^{DCD} \\ -\lambda_W^W N_v^{DCD} - \lambda_P^W Y_v^{SCS} + \lambda_V^W R_v^{DCD} & \lambda_V^V R_v^{DCD} + \lambda_P^V Y_v^{SCS} - \lambda_W^V N_v^{DCD} \end{bmatrix}$$

where $|J|$ was shown to be positive in equation (C32).

2a. Consider an increase in resource supply, $d\bar{R}^S$ positive. Then

$$\text{(C76)} \qquad \begin{aligned} dR^X &= -d\bar{R}^S < 0, \\ dN^X &= 0, \\ dY^X &= 0. \end{aligned}$$

Solving for the two endogenous variables, we obtain

$$\text{(C77)} \qquad dv^* = \lambda_V(\lambda_W(1 - (j+k)N_w^{DCD} + \lambda_P(1-m)Y_w^{SCS})d\bar{R}^S/(1-jm-kn)|J|,$$

which is negative, and

$$\text{(C78)} \qquad dw^* = -\lambda_V(\lambda_W(1 - (j+k)N_v^{DCD} + \lambda_P(1-m)Y_v^{SCS})d\bar{R}^S/(1-jm-kn)|J|,$$

which is of either sign, depending on the relative magnitudes of the exogenous variables. Thus, v* will always fall, and w* will fall or rise in response to an increase in $\bar{R}^S$.

2b. Consider an increase in autonomous demand for output, $d\bar{Y}$ positive. Then

$$\text{(C79)} \qquad \begin{aligned} dR^X &= R^{DCD}_{\bar{Y}}\, d\bar{Y} > 0, \\ dN^X &= N^{DCD}_{\bar{Y}}\, d\bar{Y} > 0, \\ dY^X &= d\bar{Y} > 0. \end{aligned}$$

Solving for the two endogenous variables, we find that v* and w* can both rise or fall, depending on the relevant magnitudes of the exogenous variables.

2c. If we perform the tedious manipulations for calculation of dv* and dw* both for neutral and for resource-augmenting technical change, we find that the signs of dv* and dw* are inconclusive, depending on the relative magnitudes of the exogenous variables.

3. Consider quasi-equilibrium Q^{NRC}_{AF}. The three excess effective demands, from Table 4.1, are

$$\text{(C80)} \qquad \begin{aligned} R^X &= R^{NCD}(v, \bar{N}^S) - \bar{R}^S > 0, \\ N^X &= N^{RCD}(w, \bar{R}^S) - \bar{N}^S > 0, \\ Y^X &= s(\bar{Y} - Y^{NRCS}(\bar{N}^S, \bar{R}^S)) > 0. \end{aligned}$$

Then

(C81) $$J^{-1} = \frac{1}{|J|}\begin{bmatrix} \lambda_W^W N_w^{RCD} & -\lambda_W^V N_w^{RCD} \\ -\lambda_V^W R_v^{NCD} & \lambda_V^V R_v^{NCD} \end{bmatrix},$$

where $|J|$ was shown to be positive in equation (C34).

3a. Consider an increase in resource supply, $d\bar{R}^S$ positive. Then

(C82) $$dR^X = -d\bar{R}^S < 0,$$

$$dN^X = N_{\bar{R}^S}^{RCD}\, d\bar{R}^S = -F_{NR}\, d\bar{R}^S/F_{NN} > 0,$$

$$dY^X = -sY_{\bar{R}^S}^{NRCS}\, d\bar{R}^S = -sF_R\, d\bar{R}^S < 0.$$

Solving for the two endogenous variables, we obtain

(C83) $$dv^* = -\lambda_W N_w^{RCD}(\lambda_V(1-(j+k))dR^X - s\lambda_P(1-n)dY^X)/(1-jm-kn)|J|$$

$$\gtrless 0$$

(C84) $$dw^* = -\lambda_V R_v^{NCD}(\lambda_W(1-(j+k))dN^X - s\lambda_P(1-m)dY^X)/(1-jm-kn)|J|,$$

which is positive. Thus we see that w* will rise, while v* may rise or fall, depending on the relative magnitudes of the exogenous variables.

3b. Consider an increase in autonomous demand for output, $d\bar{Y}$ positive. Then

(C85) $$dR^X = 0,$$

$$dN^X = 0,$$

$$dY^X = sd\bar{Y} > 0.$$

Solving for the two endogenous variables, we obtain

$$\text{(C86)} \qquad dv^* = s\lambda_W\lambda_P(1-n)\, N_w^{RCD}\, dY^X/(1-jm-kn)|J| < 0,$$

$$dw^* = s\lambda_V\lambda_P(1-m)\, R_v^{NCD}\, dY^X/(1-jm-kn)|J| < 0.$$

Thus we see that both w* and v* will fall with $d\bar{Y}$ positive.

3c. Consider neutral technical change,

$$\text{(C87)} \qquad F^2(N, R) \equiv \alpha F(N, R), \qquad \alpha > 1,$$

with $d\alpha$ positive. Then, from Appendix B5,

$$\text{(C88)} \qquad dR^X = -F_R\, d\alpha/\alpha\, F_{RR} > 0,$$

$$dN^X = -F_N\, d\alpha/\alpha\, F_{NN} > 0,$$

$$dY^X = -sF\, d\alpha < 0.$$

Solving for the two endogenous variables, we find that w* and v* both increase.

3d. Consider resource-augmenting technical change,

$$\text{(C89)} \qquad F^1(N, R) \equiv F(N, \alpha R), \qquad \alpha > 1,$$

with $d\alpha$ positive. Then, from Appendix B5,

$$\text{(C90)} \qquad dR^X = -(F_R + \alpha R^{NCD}\, F_{RR})\, d\alpha/\alpha^2\, F_{RR} \lesseqgtr 0,$$

$$dN^X = -F_{NR}\, \bar{R}^S\, d\alpha/F_{NN} > 0,$$

$$dY^X = -sF_R\, \bar{R}^S\, d\alpha < 0.$$

Solving for the two endogenous variables, we obtain

$$(C91) \qquad dv^* = \lambda_W N_w^{RCD}(\lambda_V(1 - (j+k))(F_R + \alpha R^{NCD} F_{RR})/\alpha^2 F_{RR}$$

$$- s \lambda_P(1-n)F_R \bar{R}^S) d\alpha/(1-jm-kn) |J|$$

$$\gtrless 0,$$

$$dw^* = \lambda_V R_v^{NCD}(\lambda_W(1 - (j+k)) F_{NR}/F_{NN}$$

$$- s \lambda_P(1-n) F_R) \bar{R}^S d\alpha/(1-jm-kn)|J|,$$

which is positive. Thus we see that w* will always rise, and that, for large α, v* will rise.

APPENDIX D: APPENDIX TO CHAPTER V

D1: Appendix to section 5.4.1.

The system is described by equations (5.43), which can be rewritten as

$$\text{(D1)} \qquad e = g = \lambda_V R^X - \lambda_P Y^X,$$

$$h = \lambda_W N^X - \lambda_P Y^X,$$

$$R^S = R^{SS}(e).$$

The characteristic equation for this system is

$$\text{(D2)} \qquad \begin{vmatrix} v^*g_v - Z & v^*g_w \\ w^*h_v & w^*h_w - Z \end{vmatrix} = 0.$$

From Appendix C3, we recall that the system will exhibit local dynamic stability in the region of the point (w*, v*, e*) iff the real parts of the roots of the characteristic equation are negative, $Re(Z) < 0$. It is necessary and sufficient for stability that the trace $-B$ of the Jacobian matrix J of the partial derivatives of $\dot{w}/w$ and $\dot{v}/v$ is negative and that its determinant C is positive, where, from equations (C12),

$$\text{(D3)} \qquad B = -(v^*g_v + w^*h_w),$$

$$C = w^*v^*(g_v h_w - g_w h_v) = w^*v^*|J|.$$

(i) Consider the quasi-equilibrium Q^{DC}. From equations (D1) and Tables 4.1 and 3.3, we obtain

(D4) $$g_v = \lambda_V(R_v^{DCD} - R_e^{SS} g_v) + \lambda_P Y_v^{SCS} \gtrless 0,$$

$$h_w = \lambda_W N_w^{DCD} + \lambda_P Y_w^{SCS} < 0,$$

$$g_w = \lambda_V(R_w^{DCD} - R_e^{SS} g_w) + \lambda_P Y_w^{SCS} \gtrless 0,$$

$$h_v = \lambda_W N_v^{DCD} + \lambda_P Y_v^{SCS} \gtrless 0.$$

Thus, from equations (D3), B is positive if g_v is negative, which is so, from equations (D4), iff

(D5) $$1 + \lambda_V R_e^{SS} > 0.$$

From equations (D3), C is positive if g_v is negative and $g_w h_v$ is negative, that is, if inequality (D6) holds and either

(a) $g_w < 0$ and $h_v > 0$, or

(b) $g_w > 0$ and $h_v < 0$,

as asserted in inequalities (5.47) and (5.48). From equations (D4), if inequality (D5) holds,

(D6) $$g_w \lesseqgtr 0 \quad \text{as} \quad \lambda_V/\lambda_P \lesseqgtr -Y_w^{SCS}/R_w^{DCD} > 0, \quad \text{and}$$

$$h_v \gtrless 0 \quad \text{as} \quad \lambda_W/\lambda_P \gtrless -Y_v^{SCS}/N_v^{DCD} > 0.$$

Hence we obtain the inequalities (5.49) and (5.50) as sufficient conditions for the local dynamic stability of Q^{DC}.

(ii) Consider the quasi-equilibrium Q^{NRC}. From equations (D1) and Tables 4.1 and 3.3, we obtain

$$\text{(D7)} \qquad g_v = \lambda_V(R_v^{NCD} - R_e^{SS} g_v) + s\lambda_P Y_v^{NRCS} \gtrless 0,$$

$$h_w = \lambda_W N_w^{RCD} + s\lambda_P Y_w^{NRCS} < 0,$$

$$g_w = \lambda_V(R_w^{NCD} - R_e^{SS} g_w) + s\lambda_P Y_w^{NRCS} = 0,$$

$$h_v = \lambda_W N_v^{RCD} + s\lambda_P Y_v^{NRCS} \gtrless 0.$$

But

$$\text{(D8)} \qquad Y^{NRCS} = F(\bar{N}^S, R^{SS}(e)).$$

Hence,

$$\text{(D9)} \qquad Y_v^{NRCS} = F_R R_e^{SS} g_v \gtrless 0 \quad \text{as} \quad g_v \lessgtr 0,$$

$$Y_w^{NRCS} = F_R R_e^{SS} g_w \gtrless 0 \quad \text{as} \quad g_w \lessgtr 0.$$

Substitution into equation (D7) yields

$$\text{(D10)} \qquad g_w(1 + \lambda_V R_e^{SS} - s\lambda_P F_R R_e^{SS}) = 0,$$

whence we deduce that

$$\text{(D11)} \qquad g_w = 0,$$

and the signs of equations (D7) follow.

From equations (D3) B is positive if g_v is negative, which is so, from equations (D7), if

$$\text{(D12)} \qquad 1 + \lambda_V R_e^{SS} - s\lambda_P F_R R_e^{SS} > 0.$$

From equations (D3), C is positive iff g_v is negative, hence the inequalities (5.51) and (5.52) as sufficient conditions for the local dynamic stability of Q^{NRC}. In addition,

(D13) $$B^2 - 4C = (v^*g_v - w^*h_w)^2 > 0$$

and so the quasi-equilibrium is a stable node.

D2: Appendix to section 5.5.1.

Consider the system described by equations (5.66),

(D14) $$\dot{e} = \mu\cdot(g - e),$$

$$g = \lambda_V(R^D(v) - R^{SS}(e)). \qquad \mu > 0,$$

The characteristic equation of the system is

(D15) $$\begin{vmatrix} \mu(g_e - 1) - Z & \mu g_v \\ v^*g_e & v^*g_v - Z \end{vmatrix} = 0.$$

It is necessary and sufficient for local dynamic stability that both B and C are positive, where B and C are given by

(D16) $$B = -(\mu(g_e - 1) + v^*g_v),$$

$$C = -\mu v^*g_v.$$

But, from (D14),

(D17) $$g_e = -\lambda_V R_e^{SS} > 0,$$

$$g_v = \lambda_V R_v^D \leq 0.$$

Thus we see that C is non-negative, and that

(D18) $$B = \mu \lambda_V R_e^{SS} - v^* \lambda_V R_v^D + \mu,$$

which is positive if

(D19) $$-R_e^{SS} < 1/\lambda_V - v^* R_v^D/\mu,$$

which is inequality (5.69).

Consider the three dimensional system described by equations (5.57)

(D20) $$\begin{aligned} \dot{e} &= \mu \cdot (g - e), \qquad \mu > 0, \\ g &= \lambda_V R^X - \lambda_P Y^X, \\ h &= \lambda_W N^X - \lambda_P Y^X, \\ R^S &= R^{SS}(e). \end{aligned}$$

From Birkhoff and MacLane (1977, p. 123), the system is locally dynamically stable iff all three roots of the characteristic equation

(D21) $$Z^3 + AZ^2 + BZ + C = 0$$

have negative real parts, which will occur iff the four functions A, B, C, and (AB - C) are positive when evaluated at the quasi-equilibrium. In this case, these functions are given by

(D22) $$\begin{aligned} A &= -v^*g_v - w^*h_w - \mu(g_e - 1) \\ B &= w^*v^*(g_v h_w - g_w h_v) + w^*\mu(g_e h_w - g_w h_e) - \mu(v^*g_v + w^*h_w) \\ C &= w^*v^*\mu(g_v h_w - g_w h_v). \end{aligned}$$

(i) Consider the quasi-equilibrium Q^{DC}. From equations (D20) and Tables 3.3, 4.1, and 4.3, we obtain

$$\text{(D23)} \qquad g_v = \lambda_V R_v^X - \lambda_P Y_v^X = \lambda_V R_v^{DCD} + \lambda_P Y_v^{SCS} < 0,$$

$$g_w = \lambda_V R_w^X - \lambda_P Y_w^X = \lambda_V R_w^{DCD} + \lambda_P Y_w^{SCS} \gtrless 0,$$

$$g_e = \lambda_V R_e^X - \lambda_P Y_e^X = -\lambda_V R_e^{SS} > 0,$$

$$h_w = \lambda_W N_w^X - \lambda_P Y_w^X = \lambda_W N_w^{DCD} + \lambda_P Y_w^{SCS} < 0,$$

$$h_v = \lambda_W N_v^X - \lambda_P Y_v^X = \lambda_W N_v^{DCD} + \lambda_P Y_v^{SCS} \gtrless 0,$$

$$h_e = \lambda_W N_e^X - \lambda_P Y_e^X = 0.$$

From equations (D22) and (D23), A is positive if

$$\text{(D24)} \qquad 1 + \lambda_V R_e^{SS} > 0,$$

the sufficiency condition found in Appendix D1, inequality (D5). From equation (C17), for the DC region,

$$\text{(D25)} \qquad R_v^X N_w^X = R_w^X N_v^X.$$

Thus B can be written, from equations (D22), (D23), and (D25),

$$\text{(D26)} \qquad B = w^*v^*(\lambda_W \lambda_P(N_v^X Y_w^X - N_w^X Y_v^X) + \lambda_V \lambda_P(R_w^X Y_v^X - R_v^X Y_w^X))$$

$$- w^* \mu h_w(1 + \lambda_V R_e^{SS})$$

$$- v^* \mu g_v.$$

Thus we see from Table 4.3 that B is positive if inequality (D24) holds. Using equation (D25) and Table 4.3, C is easily shown to be positive.

From equations (D22), we can write

(D27) $$A = D + \mu(E + 1),$$

$$B = F + \mu D + \mu EG,$$

$$C = \mu F,$$

where

(D28) $$D = -w^*h_w - v^*g_v > 0,$$

$$E = \lambda_V R_e^{SS} < 0,$$

$$F = w^*v^*(g_v h_w - g_w h_v) > 0,$$

$$G = -w^*h_w > 0.$$

Then

(D29) $$AB - C = \mu^2(E+1)(D + EG) + \mu(D^2 + DEG + EF) + DF,$$

which can be shown to be positive for all $\mu \geq 0$ if inequality (D24) holds, since each coefficient of $P(\mu)$ can be shown to be positive.

(ii) Consider the quasi-equilibrium Q^{NRC}. From equations (D20) and Tables 3.3, 4.1, and 4.3, we obtain

(D30) $$g_v = \lambda_V R_v^X - \lambda_P Y_v^X = \lambda_V R_v^{NCD} < 0$$

$$g_w = \lambda_V R_w^X - \lambda_P Y_w^X = 0$$

$$g_e = \lambda_V R_e^X - \lambda_P Y_e^X = -\lambda_V R_e^{SS} + s\lambda_P Y_e^{NRCS} \gtrless 0$$

$$h_w = \lambda_W N_w^X - \lambda_P Y_w^X = \lambda_W N_w^{RCD} < 0$$

$$h_v = \lambda_W N_v^X - \lambda_P Y_v^X = 0$$

$$h_e = \lambda_W N_e^X - \lambda_P Y_e^X = s\lambda_P Y_e^{NRCS} < 0,$$

where

$$\text{(D31)} \qquad Y_e^{NRCS} = F_R \, R_e^{SS} < 0.$$

From equations (D22) and (D30), A is positive if

$$\text{(D32)} \qquad 1 + \lambda_V \, R_e^{SS} - s\lambda_P \, F_R \, R_e^{SS} > 0,$$

the sufficiency condition previously found in Appendix D1, inequality (D12). B can be written, from equations (D22) and (D30),

$$\text{(D33)} \qquad \begin{aligned} B = {} & w^*v^* \, \lambda_V\lambda_W \, R_v^{NCD} \, N_w^{RCD} \\ & + w^*\mu \, \lambda_W(s\lambda_P \, F_R - \lambda_V) \, N_w^{RCD} \, R_e^{SS} \\ & - \mu(v^* \, \lambda_V \, R_v^{NCD} + w^* \, \lambda_W \, N_w^{RCD}), \end{aligned}$$

and we see that B is positive if

$$\text{(D34)} \qquad \lambda_V/\lambda_P < s \, F_R.$$

From equations (D22) and (D30), C is positive.

From equations (D22) and (D30), we can write

$$\text{(D35)} \qquad \begin{aligned} A &= D + \mu(E + 1), \\ B &= F + \mu EG + \mu D, \\ C &= \mu F, \end{aligned}$$

where

$$\text{(D36)} \qquad \begin{aligned} D &= -w^* \, \lambda_W \, N_w^{RCD} - v^* \, \lambda_V \, R_v^{NCD} > 0, \\ E &= R_e^{SS}(\lambda_V - s\lambda_P \, F_R) \gtrless 0, \\ F &= w^*v^* \, \lambda_V\lambda_W \, R_w^{NCD} \, N_w^{RCD} > 0, \\ G &= - w^* \, \lambda_W \, N_w^{RCD} > 0. \end{aligned}$$

Then

$$\text{(D37)} \qquad AB - C = \mu^2(E + 1)(D + EG) + \mu(D^2 + DEG + EF) + DF.$$

If inequalities (D32) and (D34) hold, then it can be shown that equation (D37) is positive for all non-negative μ, since each coefficient of $P(\mu)$ can be shown to be positive.

D3: Appendix to section 5.5.2.

Consider the three-dimensional system described by equations (5.74)

$$\text{(D38)} \qquad \dot{e} = \mu\cdot(g - e), \qquad \mu > 0,$$

$$g = \lambda_V R^X - \lambda_P Y^X,$$

$$h = \lambda_W N^X - \lambda_P Y^X,$$

$$R^S = R^{HS}(e).$$

Then the coefficients of the characteristic equation, A, B, and C are still given by equations (D22). Since the quasi-equilibrium occurs with constant v and w, g and h must be zero, and so e too from Figure 5.7c. From Figure 5.2, $R^{HS}(0) = R_2^S$, and $R_e^{HS}(0) = 0$. Thus the stability conditions of the quasi-equilibria should be as derived in Appendix C3 and stated in section 4.2.1.

(i) Consider the quasi-equilibrium Q^{DC}. From equations (D38) and (D23) and Tables 3.3, 4.1, and 4.3, we obtain (D39)

$$\text{(D39)} \qquad g_v = \lambda_V R_v^{DCD} + \lambda_P Y_v^{SCS} < 0,$$

$$g_w = \lambda_V R_w^{DCD} + \lambda_P Y_w^{SCS} \gtrless 0,$$

$$g_e = 0,$$

$$h_w = \lambda_W N_w^{DCD} + \lambda_P Y_w^{SCS} < 0,$$

$$h_v = \lambda_W N_v^{DCD} + \lambda_P Y_v^{SCS} \gtrless 0,$$

$$h_e = 0.$$

From equations (D22) and (D39), A is positive. From equations (D26) and (D39), B is positive. From equations (D22) and (D25), and Table 4.3, C is positive. From equations (D27), (D28), and (D29), we can write

$$\text{(D40)} \qquad AB - C = D(\mu^2 + \mu D + F), \text{ where}$$

$$D = -w^*h_w - v^*g_v > 0,$$

$$F = w^*v^*(g_v h_w - g_w h_v) > 0.$$

Thus, since each coefficient of $P(\mu)$ is positive, AB - C is positive for all non-negative μ. As expected, the quasi-equilibrium is locally stable. From the analysis in Appendix C3, if the production function is Cobb-Douglas with decreasing returns to scale, the quasi-equilibrium is a stable node.

(ii) Consider the quasi-equilibrium Q^{NRC}. From equations (D38) and (D30) and Tables 3.3, 4.1, and 4.3, we obtain

$$\text{(D41)} \qquad g_v = \lambda_V R_v^{NCD} < 0$$

$$g_w = 0$$

$$g_e = 0$$

$$h_w = \lambda_W N_w^{RCD} < 0$$

$$h_v = 0$$

$$h_e = 0.$$

From equations (D22) and (D41), A is positive. From equations (D22) and (D41), B is positive. From equations (D22) and (D41), C is positive. From equations (D35), (D36), and (D37), we can write

$$\text{(D42)} \qquad AB - C = D(\mu^2 + \mu D + F), \text{ where}$$

$$D = -w^* \lambda_W N_w^{RCD} - v^* \lambda_V R_v^{NCD} > 0,$$

$$F = w^* v^* \lambda_V \lambda_W R_v^{NCD} N_w^{RCD} > 0.$$

Thus, since each coefficient of $P(\mu)$ is positive, $AB - C$ is positive for all non-negative μ. As expected, the quasi-equilibrium is locally stable. From the analysis in Appendix C3, it is a stable node.

D4: <u>Appendix to section 5.6.1.</u>

Consider the system described by equations (5.81)

$$\text{(D43)} \qquad \dot{e} = (\alpha - e)g - \beta e$$

$$g = \lambda_V(R^D(v) - R^{SS}(e)).$$

From Figure 5.9, the quasi-equilibrium occurs with $g^* = e^* = 0$, hence, at quasi-equilibrium,

$$\text{(D44)} \qquad \partial\dot{e}/\partial e = (\alpha - e^*)g_e - (g^* + \beta) = \alpha g_e - \beta,$$

$$\partial\dot{e}/\partial v = (\alpha - e^*)g_v = \alpha g_v,$$

$$\partial\dot{e}/\partial w = (\alpha - e^*)g_w = \alpha g_w.$$

The characteristic equation of the system is thus

$$\text{(D45)} \qquad \begin{vmatrix} \alpha g_e - \beta - Z & \alpha g_v \\ v^* g_e & v^* g_v - Z \end{vmatrix} = 0.$$

It is necessary and sufficient for local dynamic stability that both B and C are positive, where B and C are given by

$$\text{(D46)} \qquad B = -\alpha g_e + \beta - v^* g_v,$$

$$C = -\beta v^* g_v.$$

But, from (D43),

$$\text{(D47)} \qquad g_e = -\lambda_V R_e^{SS} > 0,$$

$$g_v = \lambda_V R_e^D \leq 0.$$

Thus we see that C is non-negative, and that

(D48) $$B = \alpha\lambda_V R_e^{SS} + \beta - v^* \lambda_V R_v^D,$$

which is positive iff

(D49) $$-\alpha R_e^{SS} < \beta/\lambda_V - v^* R_v^D.$$

If $\alpha > 0$ (progressive expectations) then this condition imposes a limit on the steepness of $R^{SS}(e)$. If $\alpha < 0$ (regressive expectations) then there is no limit on the steepness of $R^{SS}(e)$.

Consider the three-dimensional system of equations (5.80)

(D50) $$\dot{e} = (\alpha - e)g - \beta e,$$

$$g = \lambda_V R^X - \lambda_P Y^X,$$

$$h = \lambda_W N^X - \lambda_P Y^X,$$

$$R^S = R^{SS}(e).$$

The coefficients of the characteristic equation of the system, A, B, and C, at quasi-equilibrium are given by

(D51) $$A = -v^*g_v - w^*h_w - (\alpha g_e - \beta),$$

$$B = w^*v^*(g_v h_w - g_w h_v) + w^*\alpha(g_e h_w - g_w h_e) - \beta(v^*g_v + w^*h_w),$$

$$C = w^*v^*\beta(g_v h_w - g_w h_v).$$

For local dynamic stability it is necessary and sufficient that A, B, C, and $AB - C$ are positive when evaluated at the quasi-equilibrium.

(i) Consider the quasi-equilibrium Q^{DC}. The partials of g and h are given by equations (D23). These, together with the partials of $\dot{e}$ given in equations (D44), lead to the stability analysis.

From equations (D23), (D44), and (D51), A is positive if

(D52) $$\alpha\lambda_V R_e^{SS} + \beta > 0.$$

Using equation (C17) (equation (D25)) for the DC region,

(D53) $$R_v^X N_w^X = R_w^X N_v^X,$$

we can write B, from equations (D51), as

(D54) $$\begin{aligned} B \;=\; & w^*v^*(\lambda_W\lambda_P(N_v^X Y_w^X - N_w^X Y_v^X) + \lambda_V\lambda_P(R_w^X Y_v^X - R_v^X Y_w^X)) \\ & - w^*h_w(\beta + \alpha\lambda_V R_e^{SS}) \\ & - v^*\beta g_v. \end{aligned}$$

Thus we see from Table 4.3 that B is positive if inequality (D52) holds. Using equations (D51) and (D53) and Table 4.3, C is easily shown to be positive.

From equations (D51) we can write

(D55) $$\begin{aligned} A &= D + \alpha E + \beta, \\ B &= F + \alpha EG + \beta D, \\ C &= \beta F, \end{aligned}$$

where

(D56) $$D = -w^*h_w - v^*g_v > 0,$$

$$E = \lambda_V R_e^{SS} < 0,$$

$$F = w^*v^*(g_v h_w - g_w h_v) > 0,$$

$$G = -w^*h_w > 0.$$

Then

(D57) $$AB - C = \alpha^2 E^2 G + \alpha E\cdot(F + \beta D + DG + \beta G) + DF + \beta^2 D,$$

which is always positive for negative α (regressive expectations). It is also positive for small, positive α and large, positive α, but may be negative for α in the midrange (progressive expectations). The quasi-equilibrium Q^{DC} is stable if $\alpha < 0$, or if α is small but positive.

(ii) Consider the quasi-equilibrium Q^{NRC}. The partials of g and h are given by equations (D30). These, together with the partials of $\dot{e}$ given in equations (D44), lead to the stability analysis.

From equations (D30), (D31), (D44), and (D51), A is positive if

(D58) $$\beta + \alpha R_e^{SS}\cdot(\lambda_V - s\lambda_P F_R) > 0.$$

From equations (D51), (D30), and (D31), B can be written

(D59) $$B = w^*v^* \lambda_V\lambda_W R_v^{NCD} N_w^{RCD} - w^* \lambda_W(\beta + \alpha R_e^{SS}(\lambda_V - s\lambda_P F_R)) N_w^{RCD} - v^* \lambda_V \beta R_v^{NCD},$$

and we see that B is positive if inequality (D58) holds. If

(D60) $$\lambda_V/\lambda_P < s F_R,$$

then inequality (D58) holds if α is positive. If inequality (D60) does not hold, then inequality (D58) holds if α is negative. From equations (D51), C is positive.

From equations (D51), we can write

$$\text{(D61)} \qquad A = D + \alpha E + \beta,$$

$$B = F + \alpha EG + \alpha D,$$

$$C = \beta F,$$

where

$$\text{(D62)} \qquad D = -w^* \lambda_W N_w^{RCD} - v^* \lambda_V R_v^{NCD} > 0,$$

$$E = R_e^{SS}(\lambda_V - s \lambda_P F_R) \gtrless 0,$$

$$F = w^*v^* \lambda_V\lambda_W R_v^{NCD} N_w^{RCD} > 0,$$

$$G = -w^* \lambda_W N_w^{RCS} > 0.$$

Then

$$\text{(D63)} \qquad AB - C = \alpha^2 E^2G + \alpha E\cdot(F + \beta D + DG + \beta G) + DF + \beta^2D.$$

If inequality (D60) holds, then E is positive and AB − C is positive for α positive (progressive expectations), for small, negative α and large, negative α, but may be negative for α in the midrange (regressive expectations). Thus, if inequality (D60) holds, Q^{NRC} is stable if expectations are progressive ($\alpha > 0$); if inequality (D60) does not hold, Q^{NRC} is stable if expectations are regressive ($\alpha < 0$). These are not necessary conditions.

D5: Appendix to section 5.6.2.

Consider the three-dimensional system described by equations (5.84)

(D64) $$\dot{e} = (\alpha - e)g - \beta e,$$

$$g = \lambda_V R^X - \lambda_P Y^X,$$

$$h = \lambda_W N^X - \lambda_P Y^X,$$

$$R^S = R^{HS}(e).$$

Then the coefficients of the characteristic equation of the system, A, B, and C, at quasi-equilibrium are still given by equations (D51), and the partials of the expectation adjustment equation are still given by equations (D44). Noting that the quasi-equilibria occur with constant v and w, we see that $g^* = h^* = 0$, and from Figure 5.10 $e^* = 0$. From Figure 5.2 $R^{HS}(0) = R^S_2$, and $R^{HS}_e(0) = 0$. Thus the resource supply is perfectly inelastic at quasi-equilibrium, as is the labour supply.

(i) Consider the quasi-equilibrium Q^{DC}. The partials of g and h are given by equations (D39). These, together with the partials of $\dot{e}$ given in equations (D44), lead to the stability analysis.

From equations (D51) and (D39), A is positive. From equations (D54), (D39), and Table 4.3, B is positive. From equations (D51), (D53), and Table 4.3, C is positive. From equations (D55), (D56), and (D57) we can write

(D65) $$AB - C = D(F + \beta^2), \quad \text{where}$$

$$D = -w^*h_w - v^*g_v > 0,$$

$$F = w^*v^*(g_v h_w - g_w h_v) > 0.$$

Thus we see that Q^{DC} is locally dynamically stable for all (real) α.

(ii) Consider the quasi-equilibrium Q^{NRC}. The partials of g and h are given by equations (D41). These, together with the partials of $\dot{e}$ given in equations (D44), lead to the stability analysis.

From equations (D51) and (D41), A, B, and C are positive. From equations (D61), (D62), and (D63), we can write

$$\text{(D66)} \qquad AB - C = D(F + \beta^2), \quad \text{where}$$

$$D = -w^* \lambda_W N_w^{RCD} - v^* \lambda_V R_v^{NCD} > 0,$$

$$F = w^* v^* \lambda_V \lambda_W R_v^{NCD} N_w^{RCD} > 0.$$

Thus we see that Q^{NRC} is locally dynamically stable for all (real) α.

D6: <u>Appendix to section 5.7.1</u>.

Consider the system described by equations (5.88)

$$\text{(D67)} \qquad \dot{e} = (\mu - \gamma)g - \mu e + \gamma r,$$

$$g = \lambda_V(R^D(v) - R^{SS}(e)).$$

Differentiation of the first of these leads to

$$\text{(D68)} \qquad \partial\dot{e}/\partial e = (\mu - \gamma)g_e - \mu,$$

$$\partial\dot{e}/\partial v = (\mu - \gamma)g_v,$$

$$\partial\dot{e}/\partial w = (\mu - \gamma)g_w.$$

Comparison of equations (D68) with equations (D44) demonstrates that we can use the analysis of Appendix D4, first substituting $(\mu - \gamma)$ for α and μ for β.

Thus it is necessary and sufficient for local dynamic stability of the system (D67) that both B and C are positive, where, from equations (D46), B and C are given by

(D69) $$B = -(\mu - \gamma)g_e + \mu - v^* g_v,$$

$$C = -\mu v^* g_v.$$

Cases (g) and (h) are unstable since $\mu < 0$. From equations (D47) we see that C is non-negative, and that B can be written

(D70) $$B = (\mu - \gamma)\ \lambda_V\ R_e^{SS} + \mu - v^*\ \lambda_V\ R_v^D,$$

which is positive iff

(D71) $$-(\mu - \gamma)\ R_e^{SS} < \mu/\lambda_V - v^*\ R_v^D.$$

If $(\mu - \gamma) > 0$ (cases (d), (e), (f) in Figure 5.12), this condition imposes a limit on the steepness of $R^{SS}(e)$. If $(\mu - \gamma) < 0$ (cases (a), (b) in Figure 5.12), there is no limit. B is positive in case (c), $(\mu = \gamma)$.

Consider the three-dimensional system of equations (5.87)

(D72) $$\dot{e} = (\mu - \gamma)g - \mu e + \gamma r,$$

$$g = \lambda_V\ R^X - \lambda_P\ Y^X,$$

$$h = \lambda_W\ N^X - \lambda_P\ Y^X,$$

$$R^S = R^{SS}(e).$$

The coefficients of the characteristic equation of the system, A, B, and C, at quasi-equilibrium are given by equations (D51) with α replaced by $(\mu-\gamma)$ and β by μ. Cases (g) and (h) are unstable since $\mu < 0$.

(i) Consider the quasi-equilibrium Q^{DC}. From the arguments in Appendix D4, A is positive if (from inequality (D52))

$$\text{(D73)} \qquad (\mu-\gamma)\ \lambda_V\ R_e^{SS} + \mu > 0).$$

From equation (D54), B is positive if inequality (D73) holds. If $(\mu-\gamma) \leq 0$ (cases (a), (b), and (c) in Figure 5.12), A and B are positive. Otherwise they might be negative and the system (D72) unstable. From equations (D51) and (D53) and Table 4.3, C is positive iff μ is positive.

From equations (D55) we can write

$$\text{(D74)} \qquad \begin{aligned} A &= D + (\mu-\gamma)E + \mu, \\ B &= F + (\mu-\gamma)EG + \mu D, \\ C &= \mu F, \end{aligned}$$

where, from equations (D56), D, F, and G are positive and E is negative. Then

$$\text{(D75)} \qquad AB - C = (\mu-\gamma)^2\ E^2 G + (\mu-\gamma)\ E\cdot(F + \mu D + DG + \mu G) + DF + \mu^2 D,$$

which is always positive for $\mu < \gamma$ (cases (a) and (b) in Figure 5.12). It is also positive for small, positive $(\mu-\gamma)$ (cases (c) and (d) in Figure 5.12) and for large, positive $(\mu-\gamma)$ (case (f) in Figure 5.12), but may be negative for $(\mu-\gamma)$ in the midrange (case (e) in Figure 5.12). Thus Q^{DC} is definitely stable for cases (b) and (c), definitely unstable for cases (a) and (f), and possibly unstable for cases (d) and (e), depending on the parameters.

(ii) Consider the quasi-equilibrium Q^{NRC}. From the arguments in Appendix D4, A and B are positive if (from inequality (D58))

$$\mu + (\mu - \gamma)\, R_e^{SS}(\lambda_V - s\,\lambda_P\, F_R) > 0. \tag{D76}$$

If inequality (D77) holds and if $(\mu - \gamma) \geq 0$ (cases (c), (d), and (e)),

$$\lambda_V/\lambda_P < s\, F_R, \tag{D77}$$

then inequality (D76) holds. If inequality (D77) does not hold, and if $(\mu - \gamma) \leq 0$ (cases (a), (b), and (c)), then inequality (D76) holds. From equations (D51) and (D30), C is positive iff μ is positive.

We can write A, B, and C as equations (D74), where the coefficients D, E, F, and G are given by equations (D62): D, F, and G are positive, E is positive if inequality (D77) holds. Then AB - C can be written as equation (D75), which is positive for $(\mu - \gamma)$ positive, zero, negatively small, and negatively large (cases (f), (e), (d), (c), and (a), respectively, in Figure 5.12), but may be negative, and hence unstable, for $(\mu - \gamma)$ intermediately negative (case (b) in Figure 5.12). Thus Q^{NRC} is definitely stable for case (c), definitely unstable for cases (a) and (f), and possibly stable for cases (b), (d), and (c), depending on the parameters.

D7: Appendix to section 5.7.2.

Consider the two-dimensional system described by equations (5.92)

(D78) $$\dot{e} = (\mu - \gamma)g - \mu e + \gamma r,$$

$$g = \lambda_V(R^D(v) - R^{HS}(e)).$$

Phase diagrams of this system are plotted in Figure 5.13. Note that e* is not necessarily zero, and that when

(D79) $$0 < r-a < \gamma r/\mu = e^* < r+a$$

the resource supply at quasi-equilibrium is elastic, with

(D80) $$R_e^{HS}(e^*) = -b < 0, \qquad -a < e^* - r < a.$$

This can occur in cases (b), (c), and (d) of Figure 5.13. We can use the analysis of Appendix D6 with $R_e^S = -b$ when inequalities (D79) hold, zero otherwise.

From equations (D69) and (D47) C is non-negative, and B can be written (from equation (D70)) (but cases (g) and (h) are unstable since $\mu < 0$)

(D81) $$B = (\mu - \gamma)\lambda_V R_e^{HS}(e^*) + \mu - v^* \lambda_V R_v^D,$$

which is positive iff

(D82) $$-(\mu - \gamma) R_e^{HS}(e^*) < \mu/\lambda_V - v^* R_v^D.$$

If inequality (D79) does not hold (cases (a), (e), (f) of Figure 5.13) $R_e^{HS}(e^*)$ is zero and B is positive. If inequality (D79) holds then $R_e^{HS}(e^*)$ is negative, from equation (D80). In cases (b) and (c) of Figure 5.13

$(\mu - \gamma) \leq 0$ and, from inequality (D82), B is positive. In case (d) of Figure 5.13, B is positive iff

(D83) $$(\mu - \gamma)b < \mu/\lambda_V - v^* R_v^D.$$

Consider the three-dimensional system described by equations (5.90)

(D84) $$\dot{e} = (\mu - \gamma)g - \mu e + \gamma r,$$

$$g = \lambda_V R^X - \lambda_P Y^X,$$

$$h = \lambda_W N^X - \lambda_P Y^X,$$

$$R^S = R^{HS}(e).$$

The partials of $\dot{e}$ are given by equations (D68), and comparison of these with equations (D44) demonstrates that when inequality (D79) holds we can use the analysis of Appendices D4 and D6, substituting $(\mu - \gamma)$ for α, μ for β, and $-b$ for R_e^{SS}. When inequality (D79) does not hold $R_e^{HS}(e^*)$ is zero and we can use the analysis of Appendix D5, substituting $(\mu - \gamma)$ for α and μ for β. Cases (g) and (h) are unstable since $\mu < 0$.

(i) Consider the quasi-equilibrium Q^{DC}. From the arguments in Appendices D4 and D6, A is positive if (from inequality (D73))

(D85) $$(\mu - \gamma) \lambda_V R_e^{HS}(e^*) + \mu > 0,$$

and, from equation (D54), B is positive if inequality (D85) holds. From equations (D51) and (D53) and Table 4.3, C is positive iff μ is positive. If inequality (D79) does not hold, $R_e^{HS}(e^*)$ is zero and A and B are positive, from equation (D85). If inequality (D79) holds and

$(\mu-\gamma) \leq 0$, A and B are positive, from equation (D85). If inequality (D79) holds and $(\mu-\gamma) > 0$ (case (d) of Figure 5.13), A and B are positive if, from inequality (D85),

$$1/\lambda_V > b\cdot(1-\gamma/\mu). \tag{D86}$$

From equations (D74), we can write

$$\begin{aligned} A &= D + (\mu-\gamma)E + \mu, \\ B &= F + (\mu-\gamma)EG + \mu D, \\ C &= \mu F, \end{aligned} \tag{D87}$$

where, from equations (D56),

$$\begin{aligned} D &= -w^*h_w - v^*g_v > 0, \\ E &= \lambda_V\, R_e^{HS}(e^*) \leq 0, \\ F &= w^*v^*(g_v h_w - g_w h_v) > 0, \\ G &= -w^*h_w > 0. \end{aligned} \tag{D88}$$

Then

$$AB - C = (\mu-\gamma)^2\, E^2 G + (\mu-\gamma)E\cdot(F + \mu D + DG + \mu G) + DF + \mu^2 D. \tag{D89}$$

If inequality (D79) does not hold, $R_e^{HS}(e^*)$ is zero, and E is zero, and AB - C is positive. If inequality (D79) holds, $R_e^{HS}(e^*)$ is negative, and E is negative: if $(\mu-\gamma) \leq 0$, then AB - C is positive; if $(\mu-\gamma)$ is positive (case (d) of Figure 5.13), and $R_e^{HS}(e^*) \neq 0$, then AB - C might be negative, and the system (D84) unstable. Thus Q^{DC} is definitely stable for cases (b), (c), and (e), definitely unstable for cases (a) and (f), and possibly stable for case (d), depending on the parameters.

(ii) Consider the quasi-equilibrium Q^{NRC}. From the arguments in Appendices D4 and D6, A and B are positive if (from inequality (D76))

$$\text{(D90)} \qquad \mu + (\mu - \gamma)\cdot R_e^{HS}(e^*)\cdot(\lambda_V - s\ \lambda_P\ F_R) > 0.$$

If inequality (D79) does not hold, $R_e^{HS}(e^*)$ is zero, and A and B are positive, from inequality (D90). If inequalities (D79) and (D91) hold

$$\text{(D91)} \qquad \lambda_V/\lambda_P < s\ F_R$$

and if $(\mu - \gamma) \leq 0$, then A and B are positive. If inequality (D79) holds and if inequality (D91) does not hold and if $(\mu - \gamma) \geq 0$, then A and B are positive. Thus A and B are definitely positive for cases (c) and (e). For cases (b) and (d), they are positive if

$$\text{(D92)} \qquad \mu - (\mu - \gamma)\cdot b\cdot(\lambda_V - s\ \lambda_P\ F_R) > 0.$$

We can write A, B, and C as equations (D87), and hence AB - C as equation (D89), where, from equations (D62),

$$\text{(D93)} \qquad D = -w^*\ \lambda_W\ N_w^{RCD} - v^*\ \lambda_V\ R_v^{NCD} > 0,$$

$$E = R_e^{HS}(e^*)(\lambda_V - s\ \lambda_P\ F_R) \gtrless 0,$$

$$F = w^* v^*\ \lambda_V \lambda_W\ R_v^{NCD}\ N_w^{RCD} > 0,$$

$$G = -w^*\ \lambda_W\ N_w^{RCD} > 0.$$

If inequality (D79) does not hold, $R_e^{HS}(e^*)$ is zero, and E is zero, and

AB - C is positive. If inequalities (D79) and (D91) hold, and if $(\mu - \gamma) \leq 0$, then AB - C is positive. If inequality (D79) holds and if inequality (D91) does not hold and if $(\mu - \gamma) \geq 0$, then AB - C is positive. Thus Q^{NRC} is definitely stable for cases (c) and (e), definitely unstable for cases (a) and (f), and possibly unstable for cases (b) and (d), depending on the parameters.

D8: Appendix to section 5.8.1.

Consider the system described by equations (5.98)

(D94) $$\dot{e} = (\mu - \gamma)g - \mu e + \gamma r,$$

$$g = \lambda_V(R^D(v) - R^{SS}(e)) + e.$$

The partials of $\dot{e}$ are given by equations (D68), and the coefficients B and C of the characteristic equation by equations (D69), where

(D95) $$g_e = -\lambda_V R_e^{SS} + 1 > 0,$$

$$g = \lambda_V R_v^D \leq 0.$$

Then, from equations (D69) we see that C is non-negative except for unstable cases (g) and (h), and that B can be written

(D96) $$B = (\mu - \gamma) \lambda_V R_e^{SS} + \gamma - v^* \lambda_V R_v^D,$$

which is positive iff

(D97) $$-(\mu - \gamma) \lambda_V R_e^{SS} - \gamma < -v^* \lambda_V R_v^D.$$

If $(\mu - \gamma) \leq 0$ and $\gamma > 0$ (cases (a), (b), and (c) in Figure 5.12), B is positive. Otherwise B might be negative, and the system (D94) unstable.

Consider the three-dimensional system of equations (5.97)

(D98) $$\dot{e} = (\mu - \gamma)g - \mu e + \gamma r,$$

$$g = \lambda_V R^X - \lambda_P Y^X + e,$$

$$h = \lambda_W N^X - \lambda_P Y^X,$$

$$R^S = R^{SS}(e).$$

The coefficients of the characteristic equation of the system, A, B, and C, at quasi-equilibrium are given by

(D99) $$A = -v^*g_v - w^*h_w - (\mu - \gamma)g_e + \mu,$$

$$B = w^*v^*(g_v h_w - g_w h_v) + w^*(\mu - \gamma)\cdot(g_e h_w - g_w h_e) - \mu(v^*g_v + w^*h_w),$$

$$C = w^*v^*\mu(g_v h_w - g_w h_v),$$

since the partials of $\dot{e}$ are given by equations (D68). Cases (g) and (h) are unstable since $\mu < 0$.

(i) Consider the quasi-equilibrium Q^{DC}. From equations (D98) and (D23), we obtain

(D100) $$g_v = \lambda_V R_v^{DCD} + \lambda_P Y_v^{SCS} < 0,$$

$$g_w = \lambda_V R_w^{DCD} + \lambda_P Y_w^{SCS} \gtrless 0,$$

$$g_e = -\lambda_V R_e^{SS} + 1 > 0,$$

$$h_w = \lambda_W N_w^{DCD} + \lambda_P Y_w^{SCS} < 0,$$

$$h_v = \lambda_V N_v^{DCD} + \lambda_P Y_v^{SCS} \gtrless 0,$$

$$h_e = 0.$$

From equations (D99) and (D100), A is positive if

(D101) $$(\mu - \gamma)\ \lambda_V R_e^{SS} + \gamma > 0.$$

Using equation (C17) for the DC region,

(D102) $$R_v^X N_w^X = R_w^X N_v^X,$$

we can write B, from equations (D99), as

(D103) $$B = w{*}v{*}(\lambda_W\lambda_P(N_v^X Y_w^X - N_w^X Y_v^X) + \lambda_V\lambda_P(R_w^X Y_v^X - R_w^X Y_v^X))$$
$$- w{*}h_w(\gamma + (\mu - \gamma)\ \lambda_V R_e^{SS})$$
$$- v{*}\ \mu g_v.$$

Thus we see from Table 4.3 that B is positive if inequality (D101) holds. If $(\mu - \gamma) \leq 0$ and $\gamma > 0$ (cases (a), (b), and (c) in Figure 5.12), A and B are positive. Otherwise they might be negative, and the system (D98) unstable. Using equations (D99) and (D102) and Table 4.3, C is easily shown to be positive iff μ is positive.

From equations (D99) we can write

$$\text{(D104)} \qquad A = H + G + (\mu - \gamma)\, E + \gamma,$$

$$B = F + (\mu - \gamma)\, EG + \gamma G + \mu H,$$

$$C = \mu F,$$

where

$$\text{(D105)} \qquad E = \lambda_V\, R_e^{SS} < 0,$$

$$F = w^* v^* (g_v h_w - g_w h_v) > 0,$$

$$G = -w^* h_w > 0,$$

$$H = -v^* g_v > 0,$$

$$K = (\mu - \gamma) E + \gamma \gtrless 0.$$

Then

$$\text{(D106)} \qquad AB - C = K^2 G + K\cdot(HG + G^2 + F + \mu H) + (FH + FG + \mu H^2 + \mu(HG - F)).$$

Since $HG - F$ can easily be shown to be positive, from equations (D100), we see that if inequality (D101) is satisfied and K is positive, $AB - C$ will be positive, since all the coefficients of P(K) are positive. Thus cases (b) and (c) in Figure 5.12 are definitely stable. Cases (a) and (f) are unstable. The other cases may or may not be stable, depending on the parameters.

(ii) Consider the quasi-equilibrium Q^{NRC}. From equations (D98), (D30) and (D31), we obtain

(D106) $$g_v = \lambda_V R_v^{NCD} < 0,$$

$$g_w = 0,$$

$$g_e = -\lambda_V R_e^{SS} + s \lambda_P F_R R_e^{SS} + 1 \gtrless 0,$$

$$h_w = \lambda_W N_w^{RCD} < 0,$$

$$h_v = 0,$$

$$h_e = s \lambda_P F_R R_e^{SS} < 0.$$

From equations (D99) and (D106), A is positive if

(D107) $$(\mu - \gamma) R_e^{SS} (\lambda_V - s \lambda_P F_R) + \gamma > 0.$$

From equations (D99) and (D106) we can write B as

(D108) $$B = w^*v^* \lambda_V\lambda_W R_v^{NCD} N_w^{RCD}$$

$$+ w^*((\mu - \gamma) R_e^{SS}(-\lambda_V + s \lambda_P F_R) - \gamma) - \lambda_W N_w^{RCD}$$

$$- v^*\mu \lambda_V R_v^{NCD}$$

and we see that B is positive if inequality (D107) holds. From equations (D99) and (D106), C is positive iff μ is positive.

From equations (D99) and (D106) we can write

(D109) $$A = H + G + (\mu - \gamma)E + \gamma,$$

$$B = F + (\mu - \gamma)EG + \gamma G + \mu H,$$

$$C = \mu F,$$

where

$$\text{(D110)} \qquad E = R_e^{SS}\cdot(\lambda_V - s\,\lambda_P\,F_R) \gtrless 0,$$

$$F = w^* v^*\,\lambda_V \lambda_W\,R_v^{NCD}\,N_w^{RCD} > 0,$$

$$G = -w^*\,\lambda_W\,N_w^{RCD} > 0,$$

$$H = -v^*\,\lambda_V\,R_v^{NCD} > 0,$$

$$K = (\mu - \gamma)E + \gamma \gtrless 0.$$

Then

$$\text{(D111)} \qquad AB - C = K^2 G + K\cdot(HG + G^2 + F + \mu H) + (FH + FG + \mu H^2).$$

Thus if inequality (D107) is satisfied and K is positive, AB - C will be positive, since all the coefficients of P(K) are positive.

Consider inequality (D107). If

$$\text{(D112)} \qquad \lambda_V/\lambda_P < s\,F_R,$$

then inequality (D107) holds if $(\mu - \gamma) \geq 0$ and $\gamma > 0$ (cases (c) and (d) in Figure 5.12). If inequality (D112) is not satisfied, then inequality (D107) holds if $(\mu - \gamma) \leq 0$ and $\gamma > 0$ (cases (a), (b), and (c) in Figure 5.12). Thus case (c) is definitely stable. Cases (a) and (f) are definitely unstable. The rest may or may not be stable, depending on the parameters.

D9: Appendix to section 5.8.2.

Consider the system described by equations (5.102)

(D113) $$\dot{e} = (\mu - \gamma)g - \mu e + \gamma r$$

$$g = \lambda_V(R^D(v) - R^{HS}(e)) + e.$$

Phase diagrams of this system are plotted in Figure 5.14. As in Appendix D7, inequalities (D79), when

(D114) $$0 < r - a < e^* = \gamma r/\mu < r + a$$

the resource supply at quasi-equilibrium is elastic, with

(D115) $$R_e^{HS}(e^*) = -b, \qquad -a < e^* - r < a,$$

which can occur in cases (b), (c), and (d) of Figure 5.14. We can use the analysis of Appendix D8 with $R_e^S = -b$ when inequalities (D114) hold, zero otherwise.

From equations (D69) we see that C is non-negative except for unstable cases (g) and (h). From equations (D95) and (D96), B can be written

(D116) $$B = (\mu - \gamma)\,\lambda_V\, R_e^{HS}(e^*) + \gamma - v^*\,\lambda_V\, R_v^D,$$

which is positive iff

(D117) $$-(\mu - \gamma)\,\lambda_V\, R_e^{HS}(e^*) - \gamma < -v^*\,\lambda_V\, R_v^D.$$

If inequality (D114) does not hold and γ is positive (case (a) of Figure 5.14), $R_e^{HS}(e^*)$ is zero and B is positive. If inequality (D114) holds then $R_e^{HS}(e^*)$ is negative, from equation (D115). In cases (b) and (c) of Figure 5.14, $(\mu - \gamma) \leq 0$ and $\gamma > 0$. From inequality (D117, B is positive in these cases. In cases (d), (e), and (f) of Figure 5.14, if $R_e^{HS}(e^*) \neq 0$, B is positive iff

$$(\mathrm{D118}) \qquad (\mu - \gamma)\, \lambda_V b - \gamma < -v^* \lambda_V R_v^D.$$

Consider the three-dimensional system described by equations (5.101)

$$(\mathrm{D119}) \qquad \dot{e} = (\mu - \gamma) g - \mu e + \gamma r,$$

$$g = \lambda_V R^X - \lambda_P Y^X + e,$$

$$h = \lambda_W N^X - \lambda_P Y^X,$$

$$R^S = R^{HS}(e).$$

The partials of $\dot{e}$ are given by equations (D68). We can use the analysis of Appendix D8 with $R_e^S = -b$ when inequalities (D114) hold, zero otherwise. The coefficients of the characteristic equation of the system (D119), A, B, and C, are given by equations (D99). Cases (g) and (h) are unstable since $\mu < 0$.

(i) Consider the quasi-equilibrium Q^{DC}. The partial derivatives of g and h are given by equations (D100). From the argument in Appendix D8, A is positive if

$$(\mathrm{D120}) \qquad (\mu - \gamma)\, \lambda_V R_e^{HS}(e^*) + \gamma > 0,$$

and from equation (D103), B is positive if inequality (D120) holds. From equations (D99) and (D102), C is positive iff μ is positive. If inequality (D114) does not hold, $R_e^{HS}(e^*)$ is zero and A and B are positive from inequality (D120), if $\gamma > 0$, case (a) of Figure 5.14. If inequality (D114) holds then $R_e^{HS}(e^*)$ is negative, from equation (D115).

In cases (b) and (c) of Figure 5.14, $(\mu - \gamma) \leq 0$ and $\gamma > 0$. From inequality (D120), A and B are positive in these cases. In cases (d), (e), and (f) of Figure 5.14, if $R_e^{HS}(e^*) \neq 0$, A and B are positive if inequality (D120) holds.

We can write A, B, and C as equations (D104), and hence AB - C as equation (D106), where, from equations (D105),

(D121)
$$E = \lambda_V R_e^{HS}(e^*) \leq 0,$$

$$F = w^*v^*(g_v h_w - g_w h_v) > 0,$$

$$G = -w^*h_w > 0,$$

$$H = -v^*g_v > 0,$$

$$K = (\mu - \gamma)E + \gamma \gtrless 0.$$

Then

(D122)
$$AB - C = K^2G + K\cdot(HG + G^2 + F + \mu H) + (FH + FG + \mu H^2 + \mu(HG - F)).$$

Since HG - F can easily be shown to be positive, from equations (D100), we see that if inequality (D120) is satisfied and K is positive, AB - C will be positive, since all the coefficients of P(K) are positive. Thus cases (b) and (c) in Figure 5.14 are definitely stable. Cases (a) and (f) are unstable. The other cases may or may not be stable, depending on the parameters.

(ii) Consider the quasi-equilibrium Q^{NRC}. The partial derivatives of g and h are given by equations (D106). From the argument in Appendix D8, A and B are positive if, from inequality (D107),

(D123) $(\mu - \gamma)\ R_e^{HS}(e*)(\lambda_V - s\ \lambda_P\ F_R) + \gamma > 0.$

From equations (D99) and (D106), C is positive iff μ is positive (cases (b), (c), (d), and (e) in Figure 5.14).

We can write A, B, and C as equations (D109), and hence AB - C as, equation (D111),

(D124) $AB - C \;=\; K^2G + K\cdot(HG + G^2 + F + \mu H) + (FH + FG + \mu H^2),$

where F, G, H, and K are given by equations (D110), and

(D125) $E \;=\; R_e^{HS}(e*)(\lambda_V - s\ \lambda_P\ F_R).$

Then if inequality (D123) is satisfied and K is positive, AB - C will be positive, since all the coefficients of P(K) are positive.

Consider inequality (D123). If

(D126) $\lambda_V/\lambda_P < s\ F_R,$

and if $(\mu - \gamma) \geq 0$ and $\gamma > 0$, then inequality (D123) holds, whether or not inequality (D114) holds (cases (c) and (d) in Figure 5.14). If inequality (D112) is not satisfied, and if $(\mu - \gamma) \leq 0$ and $\gamma > 0$, then inequality (D123) holds, whether or not inequality (D114) holds (cases (a), (b), and (c) in Figure 5.14). In addition, if inequality (D114) does not hold and if γ is positive, then inequality (D123) holds. Cases (a) and (f) are definitely unstable. Case (c) is definitely stable. Cases (b), (d), and (e) may or may not be stable, depending on the parameters.

INDEX OF NAMES

For Product Safety Concerns and Information please contact our EU representative GPSR@taylorandfrancis.com
Taylor & Francis Verlag GmbH, Kaufingerstraße 24, 80331 München, Germany

www.ingramcontent.com/pod-product-compliance
Ingram Content Group UK Ltd.
Pitfield, Milton Keynes, MK11 3LW, UK
UKHW022008180926
475701UK00003B/265

* 9 7 8 1 1 3 8 0 9 0 5 5 2 *